AF779862

Recent Advances in Graphene and Other Two-Dimensional Materials

Recent Advances in Graphene and Other Two-Dimensional Materials

Guest Editors

Grazia Giuseppina Politano
John Parthenios

MDPI

Basel • Beijing • Wuhan • Barcelona • Belgrade • Novi Sad • Cluj • Manchester

Guest Editors
Grazia Giuseppina Politano
Department of Environmental Engineering
University of Calabria
Rende
Italy

John Parthenios
Institute of Chemical Engineering Sciences (ICE-HT)
Foundation for Research and Technology-Hellas (FORTH)
Patras
Greece

Editorial Office
MDPI AG
Grosspeteranlage 5
4052 Basel, Switzerland

This is a reprint of the Special Issue, published open access by the journal *Crystals* (ISSN 2073-4352), freely accessible at: https://www.mdpi.com/journal/crystals/special_issues/Y04C68W0L2.

For citation purposes, cite each article independently as indicated on the article page online and as indicated below:

Lastname, A.A.; Lastname, B.B. Article Title. *Journal Name* **Year**, *Volume Number*, Page Range.

ISBN 978-3-7258-7172-8 (Hbk)
ISBN 978-3-7258-7173-5 (PDF)
https://doi.org/10.3390/books978-3-7258-7173-5

© 2026 by the authors. Articles in this reprint are Open Access and distributed under the Creative Commons Attribution (CC BY) license. The reprint as a whole is distributed by MDPI under the terms and conditions of the Creative Commons Attribution-NonCommercial-NoDerivs (CC BY-NC-ND) license (https://creativecommons.org/licenses/by-nc-nd/4.0/).

Contents

About the Editors

Grazia Giuseppina Politano

Grazia Giuseppina Politano is a postdoctoral fellow at the University of Calabria at the Department of Environmental Engineering, focusing on light–matter interaction at the nanoscale for solar membrane distillation. She earned her master's degree in physics (cum laude) in 2016 and a PhD in "Physical, Chemical and Materials Sciences and Technologies" in 2021, with research spanning graphene-based thin films, nanostructured materials, and advanced optical characterization techniques such as spectroscopic ellipsometry and Raman spectroscopy. Her multidisciplinary expertise bridges physics, chemistry, and materials science, covering topics from two-dimensional materials and nanophotonics to functional membranes for sustainable technologies. She has been recognized with the MDPI *Crystals* Outstanding Reviewer Award in 2025. Dr. Politano has served as guest editor and reviewer for leading international journals and has authored numerous scientific publications on nanomaterials and functional graphene-based films.

John Parthenios

John Parthenios is a Principal Researcher at the Institute of Chemical Engineering Sciences (ICE-HT), Foundation for Research and Technology – Hellas (FORTH), Greece. He holds a PhD in physics from the University of Patras, Greece, with postdoctoral training in laser spectroscopy and nanomaterials. His research centers on the synthesis, characterization, and strain engineering of graphene and other 2D materials (e.g., MoS_2, WS_2), with emphasis on Raman and photoluminescence spectroscopy, optical straintronics, and the development of nanocomposites for tribological and aerospace applications. He has co-authored over 90 peer-reviewed journal articles, including high-impact publications with A. Geim and K. Novoselov, recipients of the 2010 Nobel Prize in Physics, on topics related to strain and structure–property relationships in graphene. His work has received over 7,900 citations (h-index 34) and includes five book chapters and two international patents. He served as Adjunct Faculty Member at the Hellenic Open University, Lecturer at the University of Patras, and Assistant Professor (contract) at the Technological Educational Institute of Patras. His career combines academic excellence with interdisciplinary research coordination, international peer recognition, and commitment to education.

crystals

MDPI

Editorial

Recent Advances in Graphene and Other Two-Dimensional Materials

Grazia Giuseppina Politano [1,*] and John Parthenios [2,*]

[1] Department of Environmental Engineering, University of Calabria, 87036 Rende, Italy
[2] Institute of Chemical Engineering Sciences (ICE-HT), Foundation for Research and Technology-Hellas (FORTH), 26504 Patras, Greece
* Correspondence: grazia.politano@unical.it (G.G.P.); jparthen@iceht.forth.gr (J.P.)

1. Introduction

Research on breakthroughs in the synthesis and characterization of graphene [1], graphene oxide (GO) and reduced graphene oxide (rGO) [2], Molybdenum Disulfide (MoS_2) [3], and two-dimensional (2D) topological materials [4] has emerged as a critical area due to their unique electronic, optical, and mechanical properties, which enable transformative applications in electronics, energy storage, and quantum technologies [5].

Despite remarkable progress over the past two decades, the transition from fundamental discoveries to practical applications remains challenging. It has become increasingly evident that the key obstacles facing research in graphene and 2D materials lie not merely in understanding their intrinsic properties, but in developing the capability to reliably synthesize, characterize, model, and seamlessly integrate these materials into functional architectures at scale. Critical challenges persist in several interconnected areas: the lack of predictive theoretical models that can guide experimental design; the profound complexity introduced by structural disorder, edge effects, and chemical functionalization; the difficulties in achieving large-area synthesis with controlled quality, and the integration of 2D materials into practical electronic, optoelectronic, energy, and environmental applications while maintaining their exceptional properties.

This Special Issue (SI), titled *Recent Advances in Graphene and Other Two-Dimensional Materials*, brings together cutting-edge research that directly addresses these fundamental challenges. The collection showcases innovative approaches spanning advanced theoretical modeling, novel synthesis and fabrication strategies, comprehensive spectroscopic characterization, and demonstration of functional devices and applications. By bridging the gap between fundamental understanding and technological implementation, these contributions collectively advance the field toward realizing the full potential of graphene and other 2D materials in next-generation technologies.

2. An Overview of Published Articles

This SI attracted a number of submissions on graphene and other 2D materials. Nine high-quality papers were accepted following the peer-review process.

These contributions span a broad spectrum of topics, yet they naturally converge into four coherent thematic areas that capture the current frontiers of the field:

- From Modeling to Functional Graphene Nanoribbon Devices;
- Properties and Applications of Graphene Oxide and Reduced Graphene Oxide;
- Graphene-Based Silicon Heterostructures for Optoelectronic Applications;
- Two-Dimensional Materials Beyond Graphene.

Together, these thematic clusters provide a comprehensive overview of recent advances while highlighting promising directions for future research. In the following sections, we present each thematic area and discuss the individual contributions within their broader scientific context.

3. From Modeling to Functional Graphene Nanoribbon Devices

Graphene nanoribbons (GNRs) [6] have emerged as key building blocks for plasmonic and photonic technologies, owing to their reduced dimensionality, tunable electronic structure, and strong light–matter interaction. Recent advances [7] highlight the importance of combining accurate theoretical modeling with device-oriented design strategies to fully exploit their plasmonic response across different frequency regimes. Within this context, two contributions in this SI address complementary aspects of plasmonic engineering in GNRs, spanning from fundamental modeling to functional device implementation.

The article "Modeling 2D Arrangements of Graphene Nanoribbons" (contribution 3) focuses on GNRs. Owing to their sizeable bandgaps and tunable electronic properties, GNRs represent promising platforms for photonic and plasmonic applications. While significant experimental progress has been achieved in the synthesis of GNRs, the theoretical description of their electronic and optical properties has largely relied on ab initio methods, which become impractical for wide systems due to the large number of atoms involved. The authors present a semi-analytical model based on the Dirac cone approximation to describe the electronic and plasmonic properties of widely and experimentally relevant GNRs, considering both freestanding and non-freestanding configurations.

The results show that, in the terahertz regime, the plasmon–momentum dispersion is highly sensitive to variations in ribbon width and charge carrier concentration. The model successfully reproduces the electronic properties of GNRs on Ge(001) and Au(111), as well as the plasmon spectra of graphene microribbon arrays (4 μm wide) on Si/SiO_2 and GNR arrays on Si, in good agreement with experimental observations. Finally, the potential application of GNRs for molecular sensing is discussed by considering chlorpyrifos-methyl as a test molecule, showing that the plasmon resonances of all investigated GNR systems occur at the same frequency (0.95 THz).

Main Message: Terahertz plasmon dispersion in graphene nanoribbons is highly sensitive to ribbon width and charge carrier concentration, and the model reliably matches experimental plasmon spectra across multiple substrates. These results also support GNR-based molecular sensing, with all systems showing a common resonance at ~0.95 THz using chlorpyrifos-methyl as a test molecule.

In the article "A Novel High-Performance 2-to-4 Decoder Design Utilizing a Plasmonic Well and Suspended Graphene Nanoribbon" (contribution 8), the authors present a compact 2-to-4 optical decoder based on a plasmonic well structure incorporating suspended GNRs. The device exploits the tunability of graphene chemical potential to dynamically control the propagation and confinement of surface plasmon polaritons, enabling efficient routing of optical signals toward the output ports. Finite-difference time-domain simulations are used to investigate the effects of channel geometry and graphene chemical potential on surface plasmon polariton propagation, refractive index, and transmission loss. The proposed structure, consisting of a plasmonic well 30 nm wide and 10 nm high, exhibits a low propagation loss of 0.188 dB/μm and a high figure of merit of 1950 at 40 THz. The decoder achieves a contrast ratio of 36.93 dB and crosstalk suppression of −36.93 dB, while occupying a footprint of 0.05 μm^2. These results demonstrate the high optical performance and compactness of the proposed plasmonic well-based decoder.

Main message: Tunable graphene's chemical potential enables dynamic control of surface plasmon polariton propagation, resulting in low loss, high contrast ratio, strong

https://doi.org/10.3390/cryst16020112

crosstalk suppression, and an ultra-compact footprint, demonstrating excellent potential for integrated plasmonic logic and signal-routing applications.

4. Properties and Applications of Graphene Oxide and Reduced Graphene Oxide

GO and rGO have attracted considerable attention as versatile graphene-derived materials, owing to their tunable chemical composition and surface functionality [8]. The presence of oxygen-containing groups and structural defects not only influences their optical and electronic behavior, but also enables a wide range of practical applications. In this SI, three contributions address GO- and rGO-based systems from complementary perspectives, spanning fundamental modeling, spectroscopic characterization, and application-driven studies.

The article "Modelling the Structure and Optical Properties of Reduced Graphene Oxide Produced by Laser Ablation: Insights from XPS and Time-Dependent DFT" (contribution 5) investigates graphene derivatives, specifically rGO and graphene-based composites, which are regarded as promising materials for optoelectronic and photodetection applications. Motivated by the growing interest in simple and environmentally friendly graphene production methods, the study addresses unresolved fundamental issues related to the structure of these materials and the interactions between their components.

The authors propose several atomistic models of oxidized graphene fragments based on X-ray photoelectron spectroscopy compositional analysis and density functional theory calculations, representing rGO produced by laser irradiation. Using these models, they determine the composition of oxygen-containing functional groups, their equilibrium configurations, and their influence on the electronic and optical properties of graphene sheets. The nature of the low-lying excited states and the photoactive regions is also examined. Finally, the calculated absorption spectra of the graphene sheets are compared with the experimental UV–Vis spectrum of rGO obtained by laser ablation.

Main message: The work clarifies the role of oxygen functional groups in electronic structure, low-energy excited states, and optical absorption, showing good agreement between calculated spectra and experimental UV–Vis data and providing insight into photoactive regions of rGO.

The article "Unleashing the Power of Graphene-Based Nanomaterials for Chromium(VI) Ion Elimination from Water" (contribution 6) studies GO-based nanomaterials for heavy metal removal from water. GO and graphene-oxide-coated silica nanoparticles (GO/SiO_2) are synthesized and evaluated for the removal of Cr(VI) ions. The prepared nanosorbents are characterized using FTIR, Raman spectroscopy, and transmission electron microscopy. The effects of pH and Cr(VI) ion concentration on the adsorption process are examined through batch experiments for both GO and GO/SiO_2. The results show that the highest removal efficiency occurs at pH 3 for both adsorbents. At a Cr(VI) concentration of 50 ppm, GO/SiO_2 exhibits a higher removal percentage (92.28%) compared to GO (86.15%). The study confirms that Cr(VI) removal is strongly dependent on both pH and ion concentration and demonstrates that both GO and GO/SiO_2 are effective adsorbents, with GO/SiO_2 showing enhanced performance relative to GO for Cr(VI) removal from water.

Main message: This study investigates GO and GO-coated silica nanoparticles for Cr(VI) removal from water. Batch adsorption experiments show strong pH and concentration dependence, with optimal performance at pH 3. GO/SiO_2 achieves higher removal efficiency (92.28% at 50 ppm) than GO (86.15%), demonstrating enhanced adsorption capability for water remediation applications.

The review "Recent Advances in the Raman Investigation of Structural and Optical Properties of Graphene and Other Two-Dimensional Materials" (contribution 1) summa-

https://doi.org/10.3390/cryst16020112

rizes recent progress in the investigation of graphene and other 2D materials. The review includes studies on GO thin films deposited on different substrates, such as titanium and silver thin films. Particular attention is devoted to the role of surface-enhanced Raman scattering (SERS) and to the effects of film thickness and laser power on the Raman response. Studies on a silver/GO/gold sandwich configuration reveal the appearance of sharp Raman modes associated with localized surface plasmon resonances.

Main message: Recent progress in the investigation of graphene and other 2D materials is summarized.

5. Graphene-Based Silicon Heterostructures for Optoelectronic Applications

The interaction between graphene-based materials and silicon (Si) has attracted growing interest for applications in optoelectronics, photovoltaics, and sensing [9]. A detailed understanding of charge transport, interfacial effects, and defect-related phenomena is essential to fully exploit the potential of these hybrid systems. Accordingly, two studies in this SI investigate graphene-based layers on Si substrates, providing insights into their electrical performance and transport mechanisms.

In the paper "Low-Frequency Noise Characteristics of Graphene/h-BN/Si Junctions" (contribution 7), the authors investigate graphene/h-BN/Si heterostructures, which are considered promising for infrared detection and photovoltaic applications due to their tunable electrical properties and well-matched interfaces. The near lattice matching between graphene and hexagonal boron nitride (h-BN) enables the growth of graphene layers with low defect density on h-BN substrates. The study provides a detailed analysis of the low frequency electrical noise behavior of graphene/h-BN/Si heterojunctions under both forward and reverse bias at room temperature. Graphene nanolayers are directly grown on h-BN films by microwave plasma-enhanced chemical vapor deposition, while the h-BN layers are deposited by reactive high-power impulse magnetron sputtering (HiPIMS). Four different h-BN thicknesses (1 nm, 3 nm, 5 nm, and 15 nm) are examined, together with a reference graphene/Si junction fabricated under the same conditions without the h-BN interlayer. Low-frequency noise measurements are used to identify the dominant charge transport mechanisms in the different structures. The results show that grain boundaries act as major defects leading to increased noise levels under high forward bias. A statistical analysis of the voltage noise spectral density over multiple devices, supported by Raman spectroscopy, indicates that hydrogen-related defects play a significant role in the 1/f noise observed in the linear region of the current–voltage characteristics. This work provides a detailed assessment of the influence of h-BN interlayers on low-frequency noise in graphene/Si heterojunctions.

Main message: Graphene/h-BN/Si junctions are investigated to understand how electrical noise affects their performance in optoelectronic applications. By comparing different h-BN thicknesses, the authors show that defects at grain boundaries and hydrogen-related imperfections strongly influence low-frequency noise, highlighting the role of material quality in improving device reliability.

In the paper "Cyclic Voltammetry and Impedance Measurements of Graphene Oxide Thin Films Dip-Coated on n-Type and p-Type Silicon" (contribution 2), the authors address a comparatively less explored aspect of graphene research, namely the enhancement of silicon performance through interaction with graphene-based materials. Cyclic voltammetry and electrical impedance measurements are carried out on GO dip-coated on both n-type and p-type Si substrates. The results show a significant improvement in the electrical properties of GO on n-type Si, with a marked increase in conductivity and photocurrent compared to bare n-type Si. These findings highlight the potential of GO to enhance silicon-based devices and suggest possible applications in optoelectronics, contributing to the future integration of GO within the silicon semiconductor industry.

Main message: This work highlights the potential of GO to enhance silicon-based devices and outlines possible optoelectronic applications, supporting the future integration of GO within the silicon semiconductor industry.

6. Two-Dimensional Materials Beyond Graphene

Beyond graphene, a broad class of 2D materials has emerged, exhibiting physical properties that differ markedly from their bulk counterparts and enable novel functionalities. In this SI, two contributions address two-dimensional materials beyond graphene.

The article "The Pressure Response of Bulk and Two−Dimensional MoS_2 Crystals Studied by Raman and Photoluminescence Spectroscopy: Dimensionality and Pressure Transmitting Medium Effects" (contribution 9) investigates the pressure response of bulk and 2D MoS_2 crystals, including monolayer, bilayer, and multilayer samples, directly transferred onto the diamond surface of a diamond anvil cell (DAC). Raman and photoluminescence (PL) spectroscopy are employed to probe the pressure-induced changes. For high-pressure experiments on 2D MoS_2, Daphne 7474 oil and a 4:1 methanol–ethanol mixture are alternatively used as pressure-transmitting media (PTM), while Daphne 7474 oil is also used in the case of bulk MoS_2. The authors observe distinct differences in the pressure evolution of the Raman spectral features and in the pressure coefficients of the vibrational modes between bulk and 2D MoS_2 crystals of different thicknesses, with an additional dependence on the PTM employed.

Main message: Distinct pressure-dependent vibrational and optical behaviors are observed between bulk and mono-, bi-, and multilayer MoS_2, with additional sensitivity to the pressure-transmitting medium, highlighting the combined effects of dimensionality and experimental conditions on MoS_2 under compression.

In the article "Double-Heterostructure Resonant Tunneling Transistors of Surface-Functionalized Sb and Bi Monolayer Nanoribbons" (contribution 4), the authors investigate zigzag nanoribbons derived from chemically surface-modified antimony (Sb) and bismuth (Bi) monolayers, which are two-dimensional topological materials, using first-principles electronic-structure calculations. Surface functionalization with methyl, amino, or hydroxyl groups is considered to assess their potential for applications in topological transport nanoelectronics. The presence of Dirac-point-like energy dispersion near the Fermi level confirms that the scattering-forbidden edge states in these nanoribbons provide topological conductive channels with extremely high electron mobility. Based on these properties, Sb/SbXHn/Sb and Bi/BiXHn/Bi nanoribbon double heterostructures (XHn = CH_3, NH_2, OH) are designed as resonant tunneling transistors. Their electron transport characteristics are evaluated using nonequilibrium Green's function methods combined with first-principles calculations. The calculated ballistic equilibrium conduction spectra and current–voltage characteristics demonstrate that the quantum conductance arises from resonant electron tunneling between the topological edge states of the two Sb or Bi monolayer nanoribbons through the central SbXHn or BiXHn barrier. This mechanism leads to pronounced resonant current peaks accompanied by strong negative differential conductance, highlighting the suitability of these nanoribbon heterostructures for zero-loss and ultrahigh-frequency resonant tunneling applications.

Main message: This article explores antimony and bismuth nanoribbons as promising building blocks for future nanoelectronic devices. By modifying their surfaces, the authors show that these materials can efficiently control electron flow, enabling resonant tunneling behavior with high performance.

https://doi.org/10.3390/cryst16020112

7. Conclusions

The nine papers collected in this SI exemplify the versatility and transformative potential of graphene and other 2D materials in addressing fundamental scientific questions and practical technological challenges.

Looking forward, the continued advancement of research on this topic will require sustained efforts in several key areas: the development of scalable synthesis methods that maintain high material quality; the refinement of theoretical and computational tools that can handle the complexity of real-world systems, including defects, disorder, and environmental effects; and the engineering of robust integration strategies that bridge the gap between laboratory demonstrations and commercial technologies.

The integration of experimental advancements with theoretical insights, as demonstrated throughout this SI, will undoubtedly drive future breakthroughs in the field. The diverse approaches, innovative methodologies, and significant findings presented here provide both inspiration and practical guidance for the next generation of investigations. We sincerely hope that the insights and advances reported in this SI will stimulate further research and contribute to the second Volume of the SI (https://www.mdpi.com/journal/crystals/special_issues/4824H80143, accessed on 2 February 2026).

Author Contributions: All authors contributed equally to this work. All authors have read and agreed to the published version of the manuscript.

Funding: This research received no external funding.

Acknowledgments: The guest editors would like to thank all the staff at MDPI for establishing and running this SI, as well as reviewers around the globe who spent their valuable time thoroughly reviewing and improving the articles published in this SI. We also feel grateful to all the authors from Greece, Italy, Iran, Lithuania, India, Russia, China and Ecuador for choosing this SI to publish their excellent science.

Conflicts of Interest: The authors declare no conflicts of interest.

Abbreviations

The following abbreviations are used in this manuscript:

SI	Special Issue
GO	Graphene oxide
rGO	Reduced Graphene oxide
2D	Two-dimensional
MoS_2	Molybdenum Disulfide
GNRs	Graphene Nanoribbons

List of Contributions:

1. Politano, G.G.; Versace, C. Recent Advances in the Raman Investigation of Structural and Optical Properties of Graphene and Other Two-Dimensional Materials. *Crystals* **2023**, *13*, 1357. https://doi.org/10.3390/cryst13091357.
2. Politano, G.G.; Burza, S.; Versace, C. Cyclic Voltammetry and Impedance Measurements of Graphene Oxide Thin Films Dip-Coated on n-Type and p-Type Silicon. *Crystals* **2023**, *13*, 73. https://doi.org/10.3390/cryst13010073.
3. Tene, T.; Guevara, M.; Bonilla García, N.; Borja, M.; Vacacela Gomez, C. Modeling 2D Arrangements of Graphene Nanoribbons. *Crystals* **2023**, *13*, 311. https://doi.org/10.3390/cryst13020311.
4. Gong, Q.; Zhang, G. Double-Heterostructure Resonant Tunneling Transistors of Surface-Functionalized Sb and Bi Monolayer Nanoribbons. *Crystals* **2023**, *13*, 379. https://doi.org/10.3390/cryst13030379.

5. Ershov, I.V.; Lavrentyev, A.A.; Bazhin, I.V.; Holodova, O.M.; Prutsakova, N.V.; Zhdanova, T.P.; Romanov, D.L. Modelling the Structure and Optical Properties of Reduced Graphene Oxide Produced by Laser Ablation: Insights from XPS and Time-Dependent DFT. *Crystals* **2023**, *13*, 600. https://doi.org/10.3390/cryst13040600.
6. Jibin, K.P.; Augustine, S.; Velayudhan, P.; George, J.S.; Krishnageham Sidharthan, S.; Paulose, S.V.; Thomas, S. Unleashing the Power of Graphene-Based Nanomaterials for Chromium(VI) Ion Elimination from Water. *Crystals* **2023**, *13*, 1047. https://doi.org/10.3390/cryst13071047.
7. Glemža, J.; Pliaterytė, I.; Matukas, J.; Gudaitis, R.; Vasiliauskas, A.; Jankauskas, Š.; Meškinis, Š. Low-Frequency Noise Characteristics of Graphene/h-BN/Si Junctions. *Crystals* **2025**, *15*, 747. https://doi.org/10.3390/cryst15090747.
8. Maleki, M.J.; Soroosh, M. A Novel High-Performance 2-to-4 Decoder Design Utilizing a Plasmonic Well and Suspended Graphene Nanoribbon. *Crystals* **2025**, *15*, 988. https://doi.org/10.3390/cryst15110988.
9. Sorogas, N.; Tersis, K.; Michail, A.; Ves, S.; Papagelis, K.; Christofilos, D.; Arvanitidis, J. The Pressure Response of Bulk and Two−Dimensional MoS_2 Crystals Studied by Raman and Photoluminescence Spectroscopy: Dimensionality and Pressure Transmitting Medium Effects. *Crystals* **2025**, *15*, 1056. https://doi.org/10.3390/cryst15121056.

References

1. Geim, A.K.; Novoselov, K.S. The rise of graphene. *Nat. Mater.* **2007**, *6*, 183–191. [CrossRef] [PubMed]
2. Wu, J.; Lin, H.; Moss, D.J.; Loh, K.P.; Jia, B. Graphene oxide for photonics, electronics and optoelectronics. *Nat. Rev. Chem.* **2023**, *7*, 162–183. [CrossRef] [PubMed]
3. Samy, O.; Zeng, S.; Birowosuto, M.D.; El Moutaouakil, A. A Review on MoS_2 Properties, Synthesis, Sensing Applications and Challenges. *Crystals* **2021**, *11*, 355. [CrossRef]
4. Hsu, C.-H.; Huang, Z.-Q.; Crisostomo, C.P.; Yao, L.-Z.; Chuang, F.-C.; Liu, Y.-T.; Wang, B.; Hsu, C.-H.; Lee, C.-C.; Lin, H.; et al. Two-dimensional Topological Crystalline Insulator Phase in Sb/Bi Planar Honeycomb with Tunable Dirac Gap. *Sci. Rep.* **2016**, *6*, 18993. [CrossRef] [PubMed]
5. Shanmugam, V.; Mensah, R.A.; Babu, K.; Gawusu, S.; Chanda, A.; Tu, Y.; Neisiany, R.E.; Försth, M.; Sas, G.; Das, O. A Review of the Synthesis, Properties, and Applications of 2D Materials. *Part. Part. Syst. Charact.* **2022**, *39*, 2200031. [CrossRef]
6. Wang, H.; Wang, H.S.; Ma, C.; Chen, L.; Jiang, C.; Chen, C.; Xie, X.; Li, A.-P.; Wang, X. Graphene nanoribbons for quantum electronics. *Nat. Rev. Phys.* **2021**, *3*, 791–802. [CrossRef]
7. Zakharova, O.V.; Mastalygina, E.E.; Golokhvast, K.S.; Gusev, A.A. Graphene Nanoribbons: Prospects of Application in Biomedicine and Toxicity. *Nanomaterials* **2021**, *11*, 2425. [CrossRef] [PubMed]
8. Razaq, A.; Bibi, F.; Zheng, X.; Papadakis, R.; Jafri, S.H.; Li, H. Review on Graphene-, Graphene Oxide-, Reduced Graphene Oxide-Based Flexible Composites: From Fabrication to Applications. *Materials* **2022**, *15*, 1012. [CrossRef] [PubMed]
9. Singh, Y.; Subbaraman, H. Recent Advances in Graphene-Enabled Silicon-Based High-Speed Optoelectronic Devices—A Review. *Photonics* **2023**, *10*, 1292. [CrossRef]

Disclaimer/Publisher's Note: The statements, opinions and data contained in all publications are solely those of the individual author(s) and contributor(s) and not of MDPI and/or the editor(s). MDPI and/or the editor(s) disclaim responsibility for any injury to people or property resulting from any ideas, methods, instructions or products referred to in the content.

https://doi.org/10.3390/cryst16020112

crystals

MDPI

Article

The Pressure Response of Bulk and Two−Dimensional MoS_2 Crystals Studied by Raman and Photoluminescence Spectroscopy: Dimensionality and Pressure Transmitting Medium Effects

Niki Sorogas [1], Krystallis Tersis [1], Antonios Michail [2], Sotirios Ves [1], Konstantinos Papagelis [1,2], Dimitrios Christofilos [3] and John Arvanitidis [1,*]

1 Physics Department, Aristotle University of Thessaloniki, 54124 Thessaloniki, Greece; nsorogka@physics.auth.gr (N.S.); ktersis@physics.auth.gr (K.T.); ves@auth.gr (S.V.); kpapag@physics.auth.gr (K.P.)
2 Institute of Accelerating Systems and Applications, National Kapodistrian University of Athens, 15784 Athens, Greece; antmichail@upatras.gr
3 School of Chemical Engineering & Laboratory of Physics, Faculty of Engineering, Aristotle University of Thessaloniki, 54124 Thessaloniki, Greece; christop@cheng.auth.gr
* Correspondence: jarvan@physics.auth.gr; Tel.: +30-2310998213

Abstract: The pressure response of bulk and two−dimensional (2D) MoS_2 crystals (monolayer, bilayer, and many–layered), directly transferred to the diamond surface of the diamond anvil cell (DAC) used for high−pressure application, is examined by means of Raman and photoluminescence (PL) spectroscopy. For the high–pressure experiments of 2D MoS_2, the Daphne 7474 oil and the 4:1 methanol–ethanol mixture are alternatively used as pressure-transmitting media (PTM), while the former is also used as PTM in the case of bulk MoS_2. Characteristic differences are observed in the pressure evolution of the Raman spectral profile and the pressure coefficients of the peak frequencies between the bulk and the 2D MoS_2 crystals of various thicknesses, which also depend on the PTM used. These observations, along with the pressure evolution of the PL spectrum of 2D MoS_2, are ascribed to the different stress conditions in each case (hydrostatic vs. uniaxial compression perpendicular to the MoS_2 layers), the adhesion of the 2D MoS_2 crystals on the diamond anvil, as well as the nature of the particular PTM used and its interaction with the studied system.

Keywords: MoS_2; two–dimensional (2D); Raman spectroscopy; photoluminescence spectroscopy; high pressure

1. Introduction

Molybdenum disulfide (MoS_2) is an important member of the transition metal dichalcogenides (TMDs) family with the general formula MX_2 (where M is a transition metal, such as Mo, W, Ti, Zr, Hf, Nb or Ta, and X a chalcogen, such as S, Se or Te), having a wide range of current and potential applications both in its bulk and two–dimensional (2D) variant [1–6]. In the bulk form, it exhibits different structural polytypes (trigonal–*T*, hexagonal–*H*, and rhombohedral–*R*), with the most common one being the hexagonal 2*H* due to its stability [1,2,7,8]. It consists of triple atomic layers (S–Mo–S) with strong covalent bonding between Mo and S atoms. Each Mo atom is surrounded by six S atoms in a trigonal prismatic coordination around Mo, with the S atoms in the upper layer located directly above those of the lower layer. These MoS_2 layers are stacked along the *c*–axis with weak

https://doi.org/10.3390/cryst15121056

van der Waals interlayer interactions, facilitating their easy mechanical separation into 2D sheets [1,2,7–9].

The 2D MoS_2 crystals have attracted intense scientific interest due to their remarkable mechanical and electronic properties, including a thickness–dependent energy bandgap [1,2,10–12]. Namely, the bulk and few–layered (FL) hexagonal MoS_2 have an indirect energy gap of 1.2–1.6 eV, which turns into a direct one of 1.8–1.9 eV in the case of monolayer (1L) MoS_2 [12,13]. This indirect–to–direct bandgap transition causes the dramatic photoluminescence (PL) signal increase in the visible spectral region for 1L MoS_2 [2,12,14], while the bandgap can be tuned by mechanical strain and pressure application [15–17]. Apart from being an effective technique to modify the lattice and electronic structure of bulk and 2D materials, pressure application also enables the continuous increase in the interaction between a 2D system and its environment, thus constituting a method of choice to investigate such effects [18].

Because of their extreme sensitivity and ease of application, optical spectroscopic techniques are versatile tools for scientific investigations in the 2D materials field, both at ambient and high–pressure conditions. Therefore, apart from their common use for identifying the samples thickness at ambient conditions (see Section 3.2), there are several reports in the literature during the last decade dealing with the study of the pressure response of 2D MoS_2 crystals by means of PL and/or Raman spectroscopy [16,19–29]. These reports provided different results—contradictory in some cases—regarding the pressure evolution of the PL and the Raman peak energies and their interpretation. This is mainly due to the different conditions of the studied 2D crystals inside the diamond anvil cells (DACs) used for high pressure application, deposited on different substrates, usually on thinned SiO_2/Si, or even directly on the diamond anvil of the DAC (to eliminate the contribution of the additional substrate in the pressure response of the crystals) and different fluids as pressure transmitting media (PTM). Moreover, in some cases, the reported pressure coefficients of the Raman peak frequencies were extracted from the corresponding frequency vs. pressure data up to very high pressures (20–30 GPa), assuming a linear dependence. Additionally, for some of the PTMs used, the deviation from linearity is further enhanced because of the solidification of the PTM at much lower pressures, causing non-hydrostatic conditions inside the DAC.

Here, we report on a systematic high–pressure study of bulk and 2D MoS_2 crystals of various thicknesses, deposited directly on the diamond anvil of the DAC (the so–called 'substrate–free' 2D crystals), by means of Raman spectroscopy, alternatively using two different fluids as PTM in the case of the 2D crystals. In contrast to previous works that often extended to much higher pressures, our analysis is restricted to the low–to–moderate pressure regime, using only data collected before the solidification of the PTM and before the onset of structural/electronic transitions, so as to obtain reliable pressure coefficients and to unveil the effects of dimensionality, 2D crystal–substrate adhesion and PTM properties on the Raman response. The pressure evolution of the PL peaks of monolayer and bilayer MoS_2 is also examined under carefully controlled excitation conditions, with the laser power kept as low as possible and the signal averaged over a scanned area to minimize sample heating. Our approach reveals non–trivial layer– and PTM–dependent trends and is expected to shed light on the pressure response of 2D MoS_2 systems, contributing to a clearer interpretation of the existing literature.

2. Materials and Methods

The bulk hexagonal 2*H*–MoS_2 crystals (purity > 99.9%) used in the present work were purchased from HQ Graphene (Groningen, the Netherlands). The 2D (single– and few–layered) MoS_2 crystals were grown by chemical vapor deposition (CVD) on SiO_2/Si substrates, through the reaction between sulfur vapors and pre–deposited sodium molybdate (Na_2MoO_4) in solution at high temperature (800 °C) and atmospheric pressure. The concentration of the precursor Na_2MoO_4 solution was 10 mg·mL^{-1}, yielding a continuous MoS_2 film with single–, bi–, and few–layered domains epitaxially fabricated on the substrate. Further details can be found in the corresponding article by Michail et al. [30].

Raman and PL spectra were collected in the backscattering geometry by a LabRAM HR spectrometer equipped with a DuoScan^{TM} mapping system (HORIBA, Kyoto, Japan). The excitation laser beam (λ_{exc} = 514 nm) of a diode–pumped solid-state laser (Cobolt Fandango^{TM}, HÜBNER Photonics, Kassel, Germany) was focused on the sample by means of a 50× super long working distance objective (focusing spot diameter: ~2 μm). The elastically scattered light is eliminated by means of an appropriate edge filter. The laser power was set either to 1 mW (Raman measurements) or 10 and 100 μW (PL measurements), to minimize any laser–heating effects. To further reduce the heating of the samples by the excitation laser beam in the case of the PL measurements, the signal from the studied sample was averaged by continuously scanning a square area of 5 × 5 μm^2, taking advantage of the DuoScan^{TM} mapping system.

High pressure was generated by means of a gas membrane-type diamond anvil cell (DAC, Almax–easyLab, Diksmuide, Belgium) using a ~100 μm–thick drilled (~150 μm hole diameter) stainless steel gasket. Pressure was calibrated using the ruby fluorescence method [31,32]. To eliminate the effect of the SiO_2/Si substrate on the pressure response of the 2D MoS_2, the crystals were transferred directly onto the diamond anvil of the DAC ('substrate–free' 2D MoS_2 crystals), utilizing the 'surface–energy' assisted methodology with polydimethylsiloxane (PDMS) stamps [33]. The experimental setup and schematic representations of the diamond anvil cell and the crystalline structures of the studied 2D MoS_2 (monolayer—1L, bilayer—2L, and many–layered—mL [34,35]) are depicted in Figure S1 in the Supplementary Materials. In this study, two different fluids served as a pressure-transmitting medium to ensure hydrostaticity inside the sample chamber. Firstly, the high viscosity, non–polar Daphne 7474 oil, comprising large–sized molecules and solidifying above 3.7 GPa [36,37]. In order to exclude any sample–related strain and doping effects on the high–pressure response of the Raman spectrum, two different experiments were conducted in this case. Secondly, the low viscosity, polar 4:1 methanol–ethanol mixture, comprising small molecules with a solidification pressure of 10.4 GPa [38,39]. The choice of these two fluids was based not only on their distinct properties and hydrostatic behavior but also on their availability, ease of application, and widespread use (especially the alcohol mixture) for high−pressure experiments.

3. Results and Discussion

3.1. High-Pressure Raman Study of Bulk MoS_2

Firstly, we have examined, during the same experiment, the pressure response of the Raman spectrum of two bulk hexagonal MoS_2 crystals with slightly different orientations with respect to the incident excitation laser beam (bulk1 and bulk2 crystals in Figure 1a, where representative Raman spectra at various pressures are illustrated), utilizing Daphne 7474 oil as PTM. It is well–known that the number and the symmetry of the Raman active modes depend on the number of layers comprising the MoS_2 crystal (bulk, even– or odd–numbered few–layered and monolayer) and the corresponding point/space group symmetry [40,41]. The hexagonal bulk 2*H*–MoS_2, belonging to the $P6_3/mmc$ space group (D_{6h} point group symmetry) [41,42], has four Raman active modes at the Γ point of the Brillouin zone with symmetries E_{1g}, E_{2g} (E^1_{2g} and E^2_{2g}) and A_{1g} [43,44]. Except for the E^2_{2g} mode, which is assigned to relative displacements of two adjacent rigid MoS_2 layers (it appears at very low frequency, below 40 cm^{-1}, and is inaccessible due to the edge filter cut–off used in the present study), the other Raman modes are associated with intralayer vibrations [41,45]. The E_{1g} mode (in-plane motion of the S atoms) at ~287 cm^{-1} is symmetry forbidden in the backscattering geometry perpendicular to the MoS_2 layers and appears only in the case of different crystal orientations with respect to the incident laser beam (bulk2 crystal in Figure 1a). Hence, the Raman spectrum of bulk MoS_2 is dominated by the E^1_{2g} (in–plane motions of the S and Mo atoms to opposite directions) and the A_{1g} mode (out–of–plane motions of the S atoms) at ~384 and ~409 cm^{-1}, respectively [41,43,44]. Furthermore, a weak second-order Raman band, marked as 2LA in Figure 1a, also appears in the spectra in the frequency range 440–475 cm^{-1} [46].

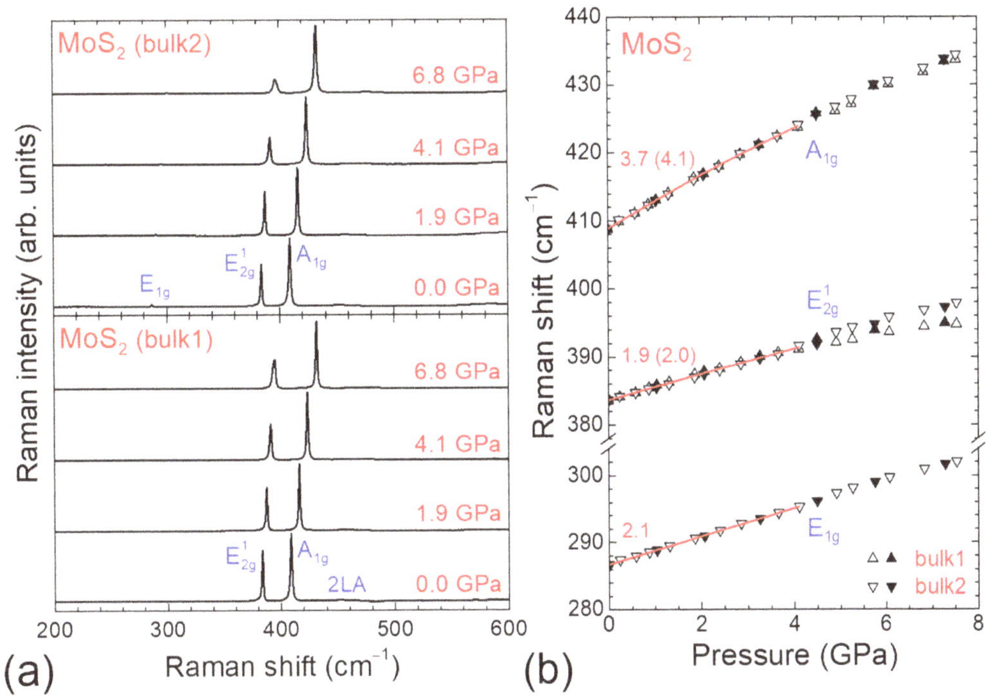

Figure 1. (**a**) Representative Raman spectra of bulk MoS_2, obtained from two crystals (bulk1 and bulk2) having slightly different orientations with respect to the incident excitation beam, at various pressures using the Daphne 7474 oil as pressure-transmitting medium (PTM). (**b**) Frequencies of the Raman active modes of the studied MoS_2 crystals as a function of pressure. Up (down) triangles represent data obtained from the bulk1 (bulk2) crystal, open (solid) symbols correspond to data obtained upon pressure increase (decrease), red lines through the experimental data are their linear or parabolic least–squares fits, while numbers refer to the linear pressure coefficients of the corresponding Raman peak frequencies. The linear pressure coefficients of the parabolic fits are given in parentheses.

With increasing pressure, the observed first-order Raman peaks shift to higher frequencies, reflecting the bond hardening upon volume reduction. At the same time, pressure application also causes the reversible gradual intensity attenuation and moderate broadening of the in–plane E^1_{2g} with respect to the out–of–plane A_{1g} Raman peak, without the appearance of any additional Raman peak (Figure 1a). The pressure evolution of the Raman peak frequencies for the two bulk MoS_2 crystals is illustrated in Figure 1b. All these pressure dependencies are quite smooth in the whole pressure range investigated (up to ~8 GPa) and fully reversible upon pressure release (solid symbols in the Figure). Note that previous

Raman studies of bulk MoS_2 up to much higher pressures have shown the appearance of a new band on the high frequency side of the in–plane mode and/or changes in the pressure coefficients for both first–order Raman peak frequencies for $P > 15$ GPa [47–50]. These observations have been attributed to the beginning of a structural phase transition from $2H_c$ to $2H_a$, caused by the interlayer sliding upon compression as suggested by the corresponding synchrotron XRD explorations for $P > 20$ GPa [51,52]. However, as expected, the experimental data presented in the current work do not show such anomalies because of the much lower pressure range investigated.

Above the solidification pressure of the Daphne 7474 oil (~4 GPa) [36], the frequency vs. pressure data is somewhat different (more pronounced for the in–plane). E^1_{2g} mode) between the two bulk crystals (up and down triangles in the Figure). This deviation reflects the gradual development of non−hydrostatic pressure components in the sample chamber inside the DAC after the solidification of the PTM used [36]. Therefore, only data obtained up to 4 GPa were used for further analysis. The linear fit of the data yields the values of 2.1, 1.9, and 3.7 $cm^{-1}GPa^{-1}$ for the pressure coefficients of the E_{1g}, E^1_{2g} and A_{1g} modes, respectively. The larger pressure coefficient for the out–of–plane mode compared to those of the in–plane ones suggests that the compression of MoS_2 on the *c*–axis is more effective compared to that of the *a*–axis [22,29]. Noticeably, the data for the main E^1_{2g} and A_{1g} Raman peaks are better fitted by parabolic (sublinear) functions, resulting in linear pressure coefficient values of 2.0 and 4.1 $cm^{-1}GPa^{-1}$, respectively. This sublinear behavior is more pronounced for the out–of–plane A_{1g} mode, reflecting the important role of the much weaker van der Waals interlayer interactions compared to the strong intralayer covalent bonds in the pressure response of bulk MoS_2. The linear pressure coefficients obtained in this work using Daphne 7474 as PTM are lying closer to the upper limit of the corresponding values in the existing literature for bulk MoS_2 using different PTM (4:1 methanol–ethanol mixture, Silicone oil, NaCl, KBr, or Ar); 1.5–1.8 $cm^{-1}GPa^{-1}$ for the E_{1g}, 1.3–1.9 $cm^{-1}GPa^{-1}$ for the E^1_{2g} and 2.6–4.0 $cm^{-1}GPa^{-1}$ for the A_{1g} mode, respectively [27,47–50,53,54].

3.2. Raman and PL Spectra of 2D MoS_2 at Ambient Pressure

Unlike bulk and even–numbered few–layer (FL) MoS_2, odd–numbered FL and 1L MoS_2 possess the $P\bar{6}m2$ space group (D_{3h} point group symmetry) due to the lack of inversion symmetry. Hence, the symmetry of the Raman active modes E_{1g}, E^1_{2g} and A_{1g} becomes E″, E′ and A_1', respectively, with identical Raman tensors because of the correlation between D_{3h} and D_{6h} [40,41,55]. Noticeably, as the number of layers in 2D MoS_2 crystals decreases, the frequency difference ($\Delta\omega$) between the out–of–plane A_{1g} (A_1') and the main in–plane E^1_{2g} (E′) mode also decreases [40,41,44,56]. This $\Delta\omega$ separation decrease is caused by the simultaneous redshift (frequency downshift) of the out–of–plane and blueshift (frequency upshift) of the in–plane mode with decreasing number of layers comprising a 2D MoS_2 crystal. The redshift of the A_{1g} (A_1') mode is attributed to the reduced restoring forces on the S atoms, resulting from the lesser interlayer interactions with decreasing number of layers [40,44]. On the other hand, it has been shown that the 'anomalous' blueshift of the E^1_{2g} (E′) mode arises from the strengthening of the surface force constants for the Mo–S intralayer interactions due to the loss of neighbors in adjacent layers in the case of 2D MoS_2 [40,57]. Apparently, the particular $\Delta\omega$ separation value can be exploited for the thickness determination of a 2D MoS_2 crystal. Figure 2a illustrates representative Raman spectra of the 'substrate–free' 2D MoS_2 sample inside the loaded DAC and prior to pressure application (nearly ambient pressure), as obtained from the three different areas shown in the optical image depicted in the inset of the Figure. The corresponding $\Delta\omega$ values are also given in the Figure. Taking into account the corresponding literature data [41,43,56], the three areas examined can be identified as many–layered (mL), bilayer (2L), and 1L MoS_2.

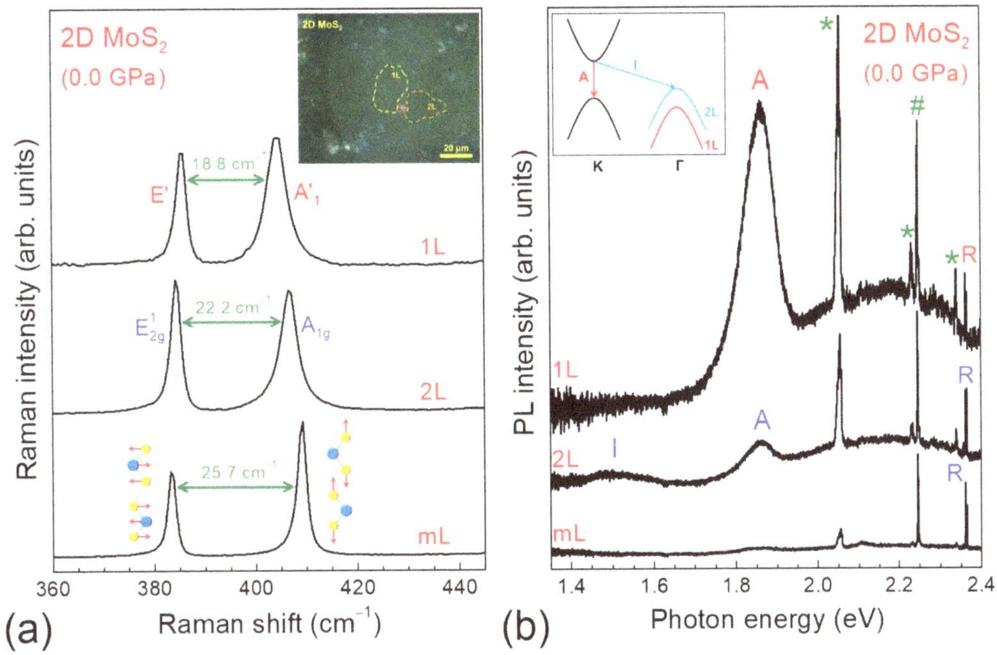

Figure 2. (**a**) Representative Raman spectra of many–layered (mL), bilayer (2L), and monolayer (1L) MoS_2 crystals inside the diamond anvil cell (DAC) prior to pressure application (0 GPa). In each case, the frequency separation ($\Delta\omega$) between the E^1_{2g} (E' for 1L) and A_{1g} (A'_1 for 1L) Raman peaks are indicated. A sketch of the E^1_{2g} and the A_{1g} phonon modes are also illustrated, where the blue (orange) circles represent the Mo (S) atoms. Inset: Optical image of the studied 1L (yellow dashed line), 2L (orange dashed line), and mL (red dashed line) crystals inside the DAC. (**b**) Typical photoluminescence (PL) spectra of mL, 2L, and 1L MoS_2 inside the DAC at nearly ambient pressure. The intensity of each spectrum was normalized to the intensity of the corresponding Raman peaks (R). Raman peaks indicated by asterisks and hash marks originate from the Daphne 7474 oil (PTM) and the diamond (DAC), respectively. Inset: Schematic diagram showing the energy band structures for 1L and 2L MoS_2 around the K and Γ points of the Brillouin zone and the interband transitions corresponding to the A and I peaks in the PL spectra.

Typical PL spectra obtained from the three aforementioned sample areas are illustrated in Figure 2b. The indirect–to–direct bandgap transition on going from 2L to 1L MoS_2 is clearly reflected in the PL spectral profile and signal intensity, being compatible with the assignment of the various sample regions to specific 2D MoS_2 crystals on the basis of the corresponding Raman spectra. In the 2L MoS_2, two rather weak bands appear in the spectrum. The lower energy broad one at ~1.52 eV, marked by 'I', is attributed to the indirect K to Γ point transition, while the higher energy band of excitonic origin at ~1.86 eV, labeled as 'A', results from the direct interband transition between the conduction band (CB) and the top of the valence band (VB) at the K point of the Brillouin zone (inset in the Figure) [19]. On the other hand, only a significantly strong excitonic A–band appears in the PL spectrum of 1L MoS_2 at a similar energy to that of the 2L and in agreement with the literature data [19,58].

3.3. High-Pressure Raman and PL Study of 2D MoS_2

Next, we studied, using again Daphne 7474 as PTM, the pressure response of the Raman spectra obtained from the three different regions of the 'substrate–free' 2D MoS_2

sample shown in the inset of Figure 2a, corresponding to 1L, 2L, and mL MoS_2. Representative Raman spectra of 1L (bottom panel) and 2L (top panel) MoS_2 crystals at various pressures up to 4 GPa, where the PTM still provides good hydrostatic conditions, are illustrated in Figure 3a. Similarly to the case of the bulk crystals, both the in–plane (E′ or E^1_{2g}) and the out–of–plane ($A_1^{'}$ or A_{1g}) Raman peaks of the 2D MoS_2 crystals gradually shift to higher frequencies with increasing pressure. Contrary to the pressure evolution of the Raman spectral profile for bulk MoS_2, pressure application on 1L and 2L crystals causes a significant broadening and peak height decrease in the out–of–plane peak, indicating the gradual deformation of the crystals upon compression. This behavior, different from that of the bulk MoS_2, is an indication of the presence of non–hydrostatic compression components on the 'substrate–free' 1L and 2L crystals and their good adhesion to the diamond surface of the DAC (vide infra).

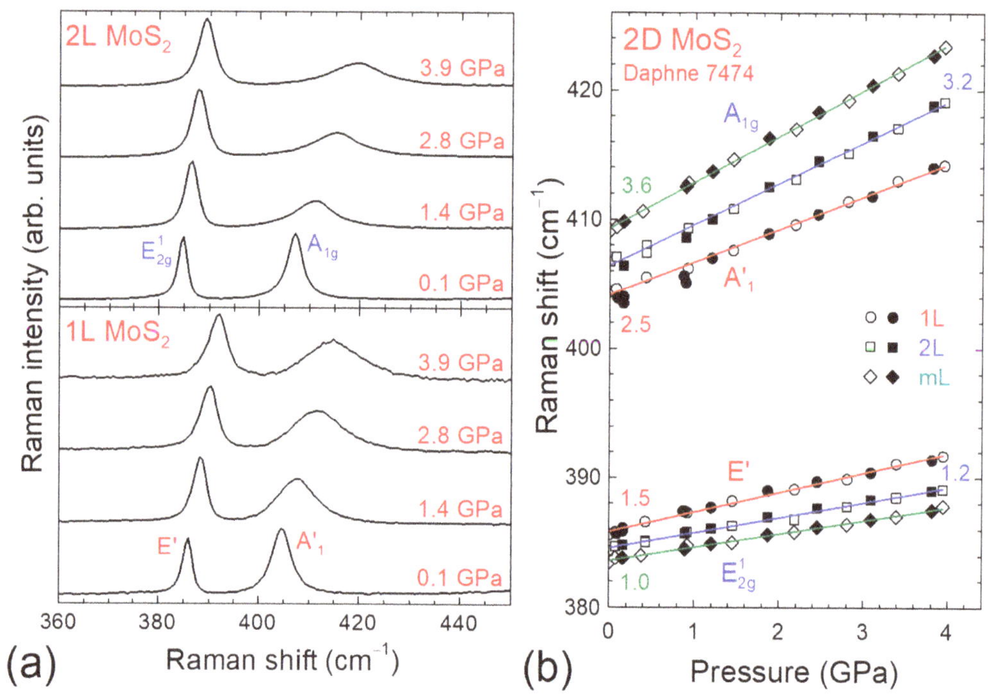

Figure 3. (**a**) Representative Raman spectra of monolayer (1L) and bilayer (2L) MoS_2 crystals at various pressures using the Daphne 7474 oil as PTM. (**b**) Frequencies of the E^1_{2g} (E′ for 1L) and A_{1g} ($A_1^{'}$ for 1L) Raman peaks of 2D MoS_2 crystals as a function of pressure. Circles, squares, and rhombi denote data obtained from 1L, 2L, and many–layered (mL) MoS_2 crystals, respectively. Open (solid) symbols correspond to data obtained upon pressure increase (decrease). Red, blue and green lines through the experimental data for 1L, 2L and mL MoS_2, respectively, are their linear least–square fits, while numbers refer to the pressure coefficients of the corresponding Raman peak frequencies in $cm^{-1}GPa^{-1}$.

As it can also be inferred from the same Figure, there is no apparent splitting of the two first–order Raman peaks up to the highest pressure attained in our experiments, contrary to the cases of uniaxial strain application along the MoS_2 layer(s), resulting in the splitting of the doubly degenerate E'/E^1_{2g} mode, or to the presence of regions with stronger and weaker layer–substrate interaction that can also cause the splitting of the $A_1^{'}/A_{1g}$ Raman peak [24,59–61]. In our case, the absence of splitting of the Raman peaks suggests that there is no significant differentiation in the in–plane strain and the MoS_2 layer–substrate interaction. Upon pressure release, only a partial recovery of the width and peak height of the out–of–plane peak in 1L and 2L MoS_2 takes place (Figure S2 in the Supplementary Materials), while pressure application has more detrimental effects and stronger irreversibility in the case of 2L MoS_2, possibly due to the additional presence of the interlayer coupling. On the other hand, pressure application on mL MoS_2 using Daphne 7474 as PTM does not have any profound effect on the overall Raman spectral profile (Figure S3 in the Supplementary Materials).

The pressure dependencies of the frequencies of the two Raman peaks observed for each of the three 'substrate–free' 2D MoS_2 crystals (1L, 2L, and mL) are illustrated in Figure 3b. All of them are quasi–linear and reversible, precluding the occurrence of any structural and/or electronic phase transition up to 4 GPa. The pressure coefficients of the Raman peak frequencies are smaller than those of the corresponding modes in the bulk crystals and increase with increasing number of layers for the out–of–plane mode (2.5, 3.2 and 3.6 $cm^{-1}GPa^{-1}$ for 1L, 2L and mL MoS_2, respectively), whereas they decrease for the in–plane mode (1.5, 1.2 and 1.0 $cm^{-1}GPa^{-1}$ for 1L, 2L and mL MoS_2, respectively). To understand the observed behavior, one should consider the stress/strain situation of the studied 2D samples inside the DAC, taking into account the nature of the PTM used, as well as the adhesion of the sample on the substrate, associated with their relative sliding [33,62].

In the case of the 2D MoS_2 crystals stuck on the diamond surface, pressure acts more or less—depending on their thickness—such as a piston along the c–axis, resulting in a predominantly uniaxial compression of the samples. Under these circumstances, the hydrostatic stress component acting on the sample (σ_H) is lower than that measured by the ruby gauge (P_R) [63,64]. Assuming pure uniaxial stress perpendicular to the sample surface with perfect adhesion on the highly stiff diamond, the hydrostatic part of the applied pressure is (Section S1 in the Supplementary Materials) [63–66]:

$$\sigma_H = \frac{(2C_{13} + C_{33})}{3C_{33}} P_R \tag{1}$$

A rough estimation of σ_H can be obtained using Equation (1), the experimental bulk MoS_2 values for the elastic constants C_{13} and C_{33} [67], keeping in mind that this is not accurate for 1L MoS_2 because of the absence of weak van der Waals interlayer interactions [33]. Remarkably, this substitution yields $\sigma_H = 0.63P_R$, in very good agreement with the reduced linear pressure coefficients for the Raman peak frequencies of 1L (Figure 3b) with respect to those of the bulk crystals (Figure 1b).

As the number of layers increases, going from 1L to 2L and eventually to mL in the 2D MoS_2 crystals, the pressure will be applied more and more hydrostatically on the samples. This leads to an increase in the σ_H/P_R ratio and could explain the gradual increase in the pressure coefficient of the out–of–plane A_{1g} mode frequency observed in Figure 3b with increasing number of layers. On the other hand, the increase in the number of layers comprising a 2D MoS_2 crystal is also expected to cause its unbinding from the substrate due to the increase in its bending modulus [24,68,69]. The reduced adhesion of the sample to the substrate promotes their relative sliding, leading to a decrease in the pressure coefficient of the in–plane E^1_{2g} mode frequency as long as pressure is applied predominantly perpendicular to the MoS_2 layers.

The pressure evolution of the PL peak energies in 1L and 2L 'substrate–free' MoS_2 crystals is also examined, reducing as much as possible the heating of the samples by the excitation laser beam (see Section 2). We stress that the pressure evolution of the PL spectrum is generally expected to be more complicated than that of the Raman peaks. Apart from the exact strain conditions of the sample in the DAC, it is affected by the different pressure dependence of the VB and the CB and their splitting, the possible electronic transitions, the presence of pressure-induced defects, and the sample heating by the excitation beam. As can be inferred from Figure 4, with 100 μW excitation power, the PL bands for both 1L and 2L shift initially with pressure (up to ~1.5 GPa) to lower energies at a rate of 13–15 meV·GPa^{-1} for the A- and ~66 meV·GPa^{-1} for the I-band, respectively. Subsequently, the position of the A-band for both 2D crystals remains nearly constant with further pressure increase up to ~4 GPa (the solidification pressure of Daphne 7474 used as PTM), while the weak I-band cannot be followed above 1.5 GPa because of its broad lineshape and its further intensity reduction with pressure.

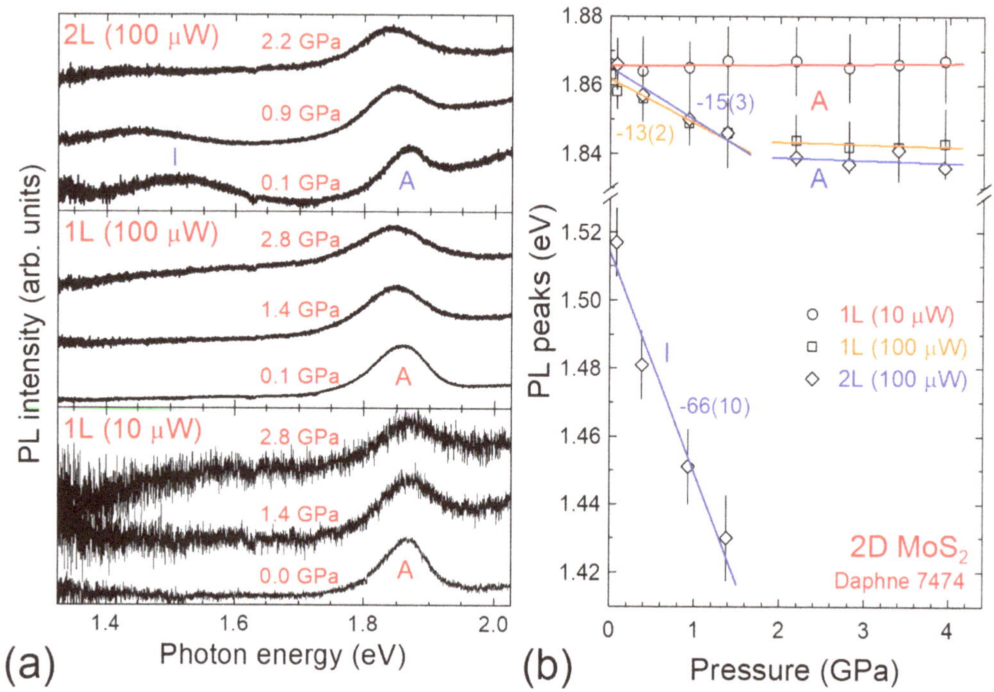

Figure 4. (**a**) Representative photoluminescence (PL) spectra of monolayer (1L) and bilayer (2L) MoS_2 crystals at various pressures using the Daphne 7474 oil as PTM. Two different laser powers, 10 (1L and 2L) and 100 μW (2L), were used for excitation. (**b**) PL peak energies as a function of pressure for 1L (A peak) and 2L (I and A peak) MoS_2. Circles, squares, and rhombi denote data obtained from 1L (10 μW), 1L (100 μW), and 2L (100 μW) MoS_2 crystals, respectively. Red, orange, and blue lines through the experimental data for 1L (10 μW), 1L (100 μW), and 2L (100 μW) MoS_2, respectively, are their linear least-squares fits, while numbers refer to the pressure coefficients of the corresponding PL peak energies in meV·GPa^{-1}.

In the case of the 2L MoS_2, the red shift in the I-band is compatible with the corresponding literature, where the decrease in the Γ–K energy gap is attributed to the large band splitting at the Γ point due to the interlayer electronic coupling upon compression [19,28,29,70]. On the other hand, the overall pressure evolution of the A-band in 2L MoS_2 deviates from the corresponding findings by Dou et al. and Li et al. [19,28]. Dou et al. used thinned SiO_2/Si substrates to support 2L MoS_2 inside the DAC, a 4:1 methanol–ethanol mixture as PTM, and much higher laser power for excitation compared to the present study (~20 times). They reported a blueshift of the A-band up to 1.5 GPa and a redshift at higher pressures, and they ascribed this discontinuity to a pressure-induced modification from a direct interband K–K to an indirect interband Λ–K transition at ~1.5 GPa [19]. On the contrary, Li et al. studied 'substrate–free' 2L MoS_2 with Ar or He as PTM and laser power ~10 times higher than the one used here [28]. They reported a continuous redshift of the A–band up to ~6 GPa at a much smaller rate (-1.6 meV·GPa^{-1}), accompanied by an intensity decrease. At higher pressures, they observed an increase in the redshift rate (–3.8 meV·GPa^{-1}), which they attributed to the same direct–to–indirect electronic transition reported by Dou et al. at ~1.5 GPa [19,28].

In the case of the 10 μW excitation power, which is more than two orders of magnitude lower than that typically used in the literature so far [16,19–21,24,25,28], the A–band for 1L MoS_2 does not practically shift with pressure in the whole pressure range investigated (0–4 GPa). Despite the special precautions taken in the present study, already using 100 μW for excitation (much lower power compared to that in the literature, and beam scanning on the sample), the different pressure response of the PL peak energy for the 10 μW laser power underlines its extreme sensitivity to the sample heating. The increase in pressure causes only the strong intensity reduction of this band, in agreement with other literature reports [19,20,25]. This intensity suppression could be ascribed to the pressure–induced intralayer hybridization between Mo *d* orbitals and the S *p* orbitals that could eventually lead to a crossover from direct to indirect transition at higher pressures and/or the pressure–induced increase in defects, as proposed in the literature [19–21,71]. The latter possibility is also compatible with the gradual deformation of the 1L and 2L MoS_2 crystals upon compression, inferred from the strong broadening and peak height decrease in the out–of–plane Raman peak (vide supra).

The essentially zero pressure shift in the A–band energy in 'substrate–free' 1L MoS_2 is in contrast to what was earlier reported in the literature for 1L MoS_2 on SiO_2/Si substrates [19–21,25] or transferred to the diamond anvil of the DAC using Ar or He as PTM [28]. In these studies, a rather strong blueshift of the A–band up to 4 GPa (the maximum pressure attained in our experiments) with rates 20–50 meV·GPa^{-1} was observed. Intriguingly, Fu et al. reported an alteration from blueshift to redshift (-15.3 meV·GPa^{-1}) above ~2 GPa for the PL peak of 1L MoS_2 on SiO_2/Si with Silicone oil as PTM, which they attributed to a K–Λ crossover in the CB [25]. The pressure–induced blueshift of the A–band in 1L MoS_2 has been ascribed to the fact that both the VB and the CB mainly originate from the $d_{x^2-y^2}$, d_{z^2} orbitals of Mo and the p_x, p_y orbitals of S, which may move away from the Fermi level by compression, leading to an increase in the energy band gap [28,72]. However, in these studies, either the substrate determines the strain of 1L MoS_2 upon compression, and hence the response of the PL peak energy [19–21,25], or the small–sized molecules of Ar and He allow for a more efficient lateral compression of the 'substrate-free' sample [28]. Our work reveals that uniaxial stress application—according to the high–pressure Raman data—up to 4 GPa in 'substrate–free' 1L MoS_2 with minimum sample heating does not practically affect the A–band energy.

The results of the present study are better correlated with the findings of Francisco–López et al., who studied the pressure response of the A and B excitons in 'substrate–free' 1L WSe_2 in comparison with that encapsulated into hexagonal boron nitride layers [33]. They observed that upon compression, the A and B exciton energies for the former sample decrease with rates -3.1 and -1.3 meV·GPa^{-1}, respectively, whereas for the latter sample, they increase with rates 3.5 and 3.8 meV·GPa^{-1}, respectively. They have ascribed these contradictory results to the essentially uniaxial stress situation, with the compressive stress component in the direction perpendicular to the plane of the 'substrate–free' WSe_2, with respect to the hydrostatic compression of the substantially thicker encapsulated sample [33]. In this context, Zhao et al. investigated the pressure response of 'substrate–free' 2D InSe crystals of various thicknesses [73]. They reported that for $N > 20$ layers of InSe, the lattice is compressed in all directions, and the intralayer compression leads to widening of the band gap, resulting in the blue shift in PL. In contrast, for $N \leq 15$, they observed a redshift of PL, which is also attributed to the predominantly uniaxial interlayer compression because of the high strain resistance along the InSe–diamond interface [73].

3.4. The Effect of the Pressure Transmitting Medium

The aforementioned findings concerning the pressure evolution of the Raman spectral profile and the pressure coefficients of the Raman peak frequencies of 'substrate–free' 2D MoS_2 crystals (1L, 2L and mL) using Daphne 7474 as PTM have been further confirmed by another, independent high–pressure experiment (Run 2) on different 2D crystals (different local strain and doping [74]). Indeed, from this second experiment, very similar results to those of the first experiment are obtained (Figures S3 and S4 in the Supplementary Materials). In order to investigate the effect of the PTM on the observed behavior, we have also studied the pressure response of 'substrate–free' 2D MoS_2 crystals by Raman spectroscopy using the 4:1 methanol–ethanol mixture as PTM. Representative Raman spectra of 1L (bottom panel) and mL (top panel) MoS_2 crystals at various pressures up to 7 GPa are illustrated in Figure 5a.

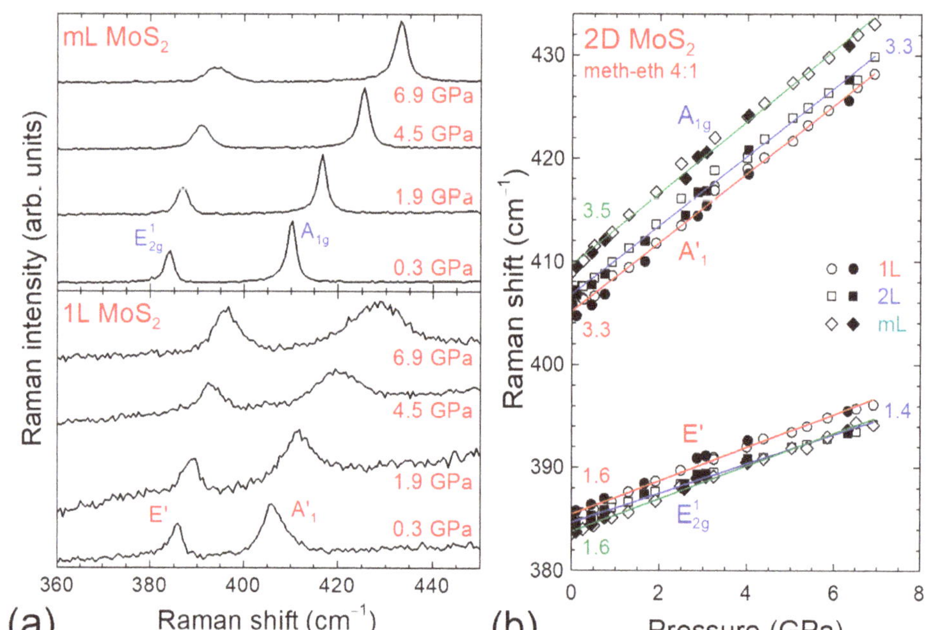

Figure 5. (**a**) Representative Raman spectra of monolayer (1L) and many–layered (mL) MoS_2 crystals at various pressures using the 4:1 methanol–ethanol mixture as PTM. (**b**) Frequencies of the E^1_{2g} (E' for 1L) and A_{1g} ($A_1^{'}$ for 1L) Raman peaks of 2D MoS_2 crystals as a function of pressure. Circles, squares, and rhombi denote data obtained from 1L, bilayer (2L), and mL MoS_2 crystals, respectively. Open (solid) symbols correspond to data obtained upon pressure increase (decrease). Red, blue and green lines through the experimental data for 1L, 2L and mL MoS_2, respectively, are their linear least–square fits, while numbers refer to the pressure coefficients of the corresponding Raman peak frequencies in $cm^{-1}GPa^{-1}$.

In this case, pressure application on 1L and 2L causes a considerably smaller broadening and relative peak height decrease in the out–of–plane $A_1^{'}$ or A_{1g} peak, while the pressure evolution of the Raman spectral profile for mL MoS_2 is very similar to that of the bulk crystals (gradual intensity attenuation and moderate broadening of the in–plane E^1_{2g} with respect to the out–of–plane A_{1g} Raman peak). Moreover, as can be inferred from Figure 5b, the pressure coefficients for the corresponding modes are similar, independently of the sample thickness (1L, 2L, or mL).

Overall, the high–pressure Raman results for the studied 2D MoS_2 samples using the alcohol mixture as PTM, the molecules of which are much smaller compared to those of Daphne 7474 [75] and can even penetrate with pressure the space between the sample and the diamond anvil [24], along with its polar nature that could lead to its stronger interaction with the samples [69], imply the more hydrostatic pressure conditions in this case. Thus, it is evident that the choice of PTM is crucial for the pressure response of 'substrate–free' 2D MoS_2 systems. Note, however, that for 2D systems on a substrate, the nature of the PTM does not appear to detectably affect its pressure response, at least in the hydrostatic pressure regime of the PTM [62]. Instead, in that case, the strain transferred from the substrate to the 2D material upon compression plays the dominant role [62,69,76].

4. Conclusions

In conclusion, we have presented our detailed high–pressure Raman and PL studies of bulk (three–dimensional, 3D) and 'substrate–free' 2D MoS_2 crystals using two different PTM in their hydrostatic pressure range. In the case of the Daphne 7474 PTM, apart from the different pressure evolution of the Raman spectral profile compared to that of the bulk, a considerable reduction in the pressure coefficients of the Raman peak frequencies in the 'substrate–free' 2D MoS_2 crystals is observed. Intriguingly, as the number of the MoS_2 layers comprising the 2D crystals increases, the pressure coefficient for the out–of–plane mode frequency increases, gradually approaching the corresponding value for the bulk, whereas for the in–plane mode it decreases on going from 1L to 2L and eventually to mL MoS_2. Furthermore, minimizing the sample heating effects, the PL peak energy in 1L MoS_2 appears to be pressure independent up to 4 GPa (close to the solidification pressure of the PTM). On the other hand, in the case of the alcohol mixture PTM, both the pressure evolution of the Raman spectral profile and the pressure coefficients of the Raman peak frequencies are more similar to those of the bulk, regardless of the number of layers comprising the 'substrate–free' 2D MoS_2 crystals. The present experimental findings and deviations can be understood in terms of the degree of uniaxial or hydrostatic stress application on going from 2D to 3D MoS_2 crystals, the degree of adhesion of the 2D crystals on the diamond surface of the high–pressure cell, as well as the properties of the PTM. It would also be interesting to extend the methodology of this study to other important—in terms of their potential applications—2D systems and/or their heterostructures, as long as 1L to FL sheets with satisfactory Raman and/or PL signal can be isolated and successfully transferred to the diamond anvil of the high–pressure cell.

Supplementary Materials: The following supporting information can be downloaded at: https://www.mdpi.com/article/10.3390/cryst15121056/s1, Figure S1: From left to right: the experimental setup, the diamond anvil cell and the crystalline structures of the studied 2D MoS_2 (monolayer—1L, bilayer—2L and many–layered—mL [34,35]); Figure S2: Representative Raman spectra of monolayer (1L) and bilayer (2L) MoS2 crystals at various pressures upon decompression and using the Daphne 7474 oil as PTM; Figure S3: Representative Raman spectra of many–layered (mL) MoS2 crystals at various pressures obtained during two different high–pressure experiments (Run one and Run two) using Daphne 7474 oil as pressure transmitting medium; Figure S4: (a) Representative Raman spectra of monolayer (1L) and bilayer (2L) MoS_2 crystals at various pressures during the second high–pressure experiment (Run 2) using the Daphne 7474 oil as pressure transmitting medium. (b) Frequencies of the E^1_{2g} (E' for 1L) and A_{1g} (A_1' for 1L) Raman peaks of 2D MoS_2 crystals as a function of pressure. Circles, squares, and rhombi denote data obtained from 1L, 2L, and many–

layered (mL) MoS_2 crystals, respectively. Red, blue, and green lines through the experimental data for 1L, 2L, and mL MoS_2, respectively, are their linear least–squares fits, while numbers refer to the pressure coefficients of the corresponding Raman peak frequencies in $cm^{-1}GPa^{-1}$; Section S1: Estimation of the hydrostatic part of the uniaxially applied stress.

Author Contributions: Conceptualization, N.S., S.V., K.P., D.C. and J.A.; methodology, N.S., A.M. and J.A.; investigation, N.S., K.T., A.M. and J.A.; formal analysis, N.S., K.T., S.V., K.P., D.C. and J.A.; writing—original draft preparation, N.S. and J.A.; writing—review and editing, S.V., K.P., D.C. and J.A.; supervision, J.A.; funding acquisition, N.S. All authors have read and agreed to the published version of the manuscript.

Funding: This research is a part of the doctoral thesis of N.S., the implementation of which was co–financed by Greece and the European Union (European Social Fund–ESF) through the Operational Program "Human Resources Development, Education and Lifelong Learning" in the context of the Act "Enhancing Human Resources Research Potential by undertaking a Doctoral Research" Sub–action 2: IKY Scholarship Program for PhD candidates in the Greek Universities.

Data Availability Statement: The original contributions presented in this study are included in the article/Supplementary Materials. Further inquiries can be directed to the corresponding author.

Acknowledgments: Support by the funding organizations is gratefully acknowledged (N.S.). The authors also acknowledge the Center of Interdisciplinary Research and Innovation of the Aristotle University of Thessaloniki (CIRI–AUTH) for access to the Raman and PL instrumentation. The research project "2DHeteroThM" implemented within the framework of the H.F.R.I call "3rd Call for H.F.R.I.'s Research Projects to Support Faculty Members and Researchers" (Lateral heterostructures of two-dimensional transition metal dichalcogenides: synthesis and thermomechanical properties, H.F.R.I. Project Number: 26296), is also acknowledged (K.P. and A.M.).

Conflicts of Interest: The authors declare no conflicts of interest.

Abbreviations

The following abbreviations are used in this manuscript:

2D	Two–dimensional
DAC	Diamond anvil cell
PL	Photoluminescence
PTM	Pressure-transmitting medium
TMD	Transition metal dichalcogenide
FL	Few–layered
1L	Monolayer
CVD	Chemical vapor deposition
PDMS	Polydimethylsiloxane
mL	Many–layered
2L	Bilayer
CB	Conduction band
VB	Valence band
3D	Three–dimensional

References

1. He, Z.; Que, W. Molybdenum disulfide nanomaterials: Structures, properties, synthesis and recent progress on hydrogen evolution reaction. *Appl. Mater. Today* **2016**, *3*, 23–56. [CrossRef]
2. Samy, O.; Zeng, S.; Birowosuto, M.D.; El Moutaouakil, A. A Review on MoS_2 properties, synthesis, sensing applications and challenges. *Crystals* **2021**, *11*, 355. [CrossRef]
3. Khaleghi, M.; Chaji, M.; Pishbin, F.; Sillanpää, M.; Sheibani, S. A review of molybdenum disulfide–based 3D printed structures for biomedical applications. *J. Mater. Res. Technol.* **2024**, *32*, 1630–1646. [CrossRef]

4. Pinto, F.M.; de Conti, M.C.M.D.; Pereira, W.S.; Sczancoski, J.C.; Medina, M.; Corradini, P.G.; de Brito, J.F.; Nogueira, A.E.; Góes, M.S.; Ferreira, O.P.; et al. Recent advances in layered MX2-based materials (M = Mo, W and X = S, Se, Te) for emerging optoelectronic and photo(electro)catalytic applications. *Catalysts* **2024**, *14*, 388. [CrossRef]
5. Dhingra, A.; Kumar, R.; Thakur, O.P.; Pandey, R. Advancements in fabrication, polymorph diversity, heterostructure excitation dynamics, and multifunctional applications of leading 2D-transition metal dichalcogenides. *J. Ind. Eng. Chem.* **2025**, *151*, 40–74. [CrossRef]
6. Alowakennu, M.; Khan, M.M.; Abdulwahab, K.O. Recent progress in photocatalytic applications of metals- and non-metals-doped MoS_2. *Mater. Sci. Semicond. Process.* **2026**, *201*, 110082. [CrossRef]
7. Benavente, E.; Santa Ana, M.A.; Mendizábal, F.; González, G. Intercalation chemistry of molybdenum disulfide. *Coord. Chem. Rev.* **2002**, *224*, 87–109. [CrossRef]
8. Krishnan, U.; Kaur, M.; Singh, K.; Kumar, M.; Kumar, A. A synoptic review of MoS_2: Synthesis to applications. *Superlattices Microstruct.* **2019**, *128*, 274–297. [CrossRef]
9. Mounet, N.; Gibertini, M.; Schwaller, P.; Campi, D.; Merkys, A.; Marrazzo, A.; Sohier, T.; Castelli, I.E.; Cepellotti, A.; Pizzi, G.; et al. Two–dimensional materials from high–throughput computational exfoliation of experimentally known compounds. *Nat. Nanotechnol.* **2018**, *13*, 246–252. [CrossRef] [PubMed]
10. Castellanos–Gomez, A.; Poot, M.; Steele, G.A.; van der Zant, H.S.J.; Agraït, N.; Rubio–Bollinger, G. Elastic properties of freely suspended MoS_2 nanosheets. *Adv. Mater.* **2012**, *24*, 772–775. [CrossRef] [PubMed]
11. Radisavljevic, B.; Radenovic, A.; Brivio, J.; Giacometti, V.; Kis, A. Single–layer MoS_2 transistors. *Nat. Nanotechnol.* **2011**, *6*, 147–150. [CrossRef]
12. Mak, K.F.; Lee, C.; Hone, J.; Shan, J.; Heinz, T.F. Atomically thin MoS_2: A new direct–gap semiconductor. *Phys. Rev. Lett.* **2010**, *105*, 136805. [CrossRef]
13. Kumar, V.P.; Panda, D.K. Review–Next generation 2D material molybdenum disulfide (MoS_2): Properties, applications and challenges. *ECS J. Solid State Sci. Technol.* **2022**, *11*, 033012. [CrossRef]
14. Venkata Subbaiah, Y.P.; Saji, K.J.; Tiwari, A. Atomically thin MoS_2: A versatile nongraphene 2D material. *Adv. Funct. Mater.* **2016**, *26*, 2046–2069. [CrossRef]
15. Blundo, E.; Cappelluti, E.; Felici, M.; Pettinari, G.; Polimeni, A. Strain–tuning of the electronic, optical, and vibrational properties of two–dimensional crystals. *Appl. Phys. Rev.* **2021**, *8*, 021318. [CrossRef]
16. Pimenta Martins, L.G.; Carvalho, B.R.; Occhialini, C.A.; Neme, N.P.; Park, J.-H.; Song, Q.; Venezuela, P.; Mazzoni, M.S.C.; Matos, M.J.S.; Kong, J.; et al. Electronic band tuning and multivalley Raman scattering in monolayer transition metal dichalcogenides at high pressures. *ACS Nano* **2022**, *16*, 8064–8075. [CrossRef] [PubMed]
17. Pimenta Martins, L.G.; Comin, R.; Matos, M.J.S.; Mazzoni, M.S.C.; Neves, B.R.A.; Yankowitz, M. High–pressure studies of atomically thin van der Waals materials. *Appl. Phys. Rev.* **2023**, *10*, 011313. [CrossRef]
18. San-Miguel, A. Nanomaterials under high–pressure. *Chem. Soc. Rev.* **2006**, *35*, 876–889. [CrossRef] [PubMed]
19. Dou, X.; Ding, K.; Jiang, D.; Sun, B. Tuning and identification of interband transitions in monolayer and bilayer molybdenum disulfide using hydrostatic pressure. *ACS Nano* **2014**, *8*, 7458–7464. [CrossRef]
20. Li, F.; Yan, Y.; Han, B.; Li, L.; Huang, X.; Yao, M.; Gong, Y.; Jin, X.; Liu, B.; Zhu, C.; et al. Pressure confinement effect in MoS_2 monolayers. *Nanoscale* **2015**, *7*, 9075–9082. [CrossRef]
21. Nayak, A.P.; Pandey, T.; Voiry, D.; Liu, J.; Moran, S.T.; Sharma, A.; Tan, C.; Chen, C.-H.; Li, L.-J.; Chhowalla, M.; et al. Pressure–dependent optical and vibrational properties of monolayer molybdenum disulfide. *Nano Lett.* **2015**, *15*, 346–353. [CrossRef]
22. Yan, Y.; Li, F.; Gong, Y.; Yao, M.; Huang, X.; Fu, X.; Han, B.; Zhou, Q.; Cui, T. Interlayer coupling affected structural stability in ultrathin MoS_2: An investigation by high pressure Raman spectroscopy. *J. Phys. Chem. C* **2016**, *120*, 24992–24998. [CrossRef]
23. Li, X.; Li, J.; Wang, K.; Wang, X.; Wang, S.; Chu, X.; Xu, M.; Fang, X.; Wei, Z.; Zhai, Y.; et al. Pressure and temperature-dependent Raman spectra of MoS_2 film. *Appl. Phys. Lett.* **2016**, *109*, 242101. [CrossRef]
24. Alencar, R.S.; Saboia, K.D.A.; Machon, D.; Montagnac, G.; Meunier, V.; Ferreira, O.P.; San-Miguel, A.; Souza Filho, A.G. Atomic-layered MoS_2 on SiO_2 under high pressure: Bimodal adhesion and biaxial strain effects. *Phys. Rev. Mater.* **2017**, *1*, 024002. [CrossRef]
25. Fu, L.; Wan, Y.; Tang, N.; Ding, Y.-M.; Gao, J.; Yu, J.; Guan, H.; Zhang, K.; Wang, W.; Zhang, C.; et al. K–Λ crossover transition in the conduction band of monolayer MoS_2 under hydrostatic pressure. *Sci. Adv.* **2017**, *3*, e1700162. [CrossRef]
26. Yang, M.; Cheng, X.; Li, Y.; Ren, Y.; Liu, M.; Qi, Z. Anharmonicity of monolayer MoS_2, $MoSe_2$, and WSe_2: A Raman study under high pressure and elevated temperature. *Appl. Phys. Lett.* **2017**, *110*, 093108. [CrossRef]
27. Cheng, X.; Li, Y.; Shang, J.; Hu, C.; Ren, Y.; Liu, M.; Qi, Z. Thickness–dependent phase transition and optical behavior of MoS_2 films under high pressure. *Nano Res.* **2018**, *11*, 855–863. [CrossRef]
28. Li, C.; Liu, Y.; Yang, Q.; Zheng, Q.; Yan, Z.; Han, J.; Lin, J.; Wang, S.; Qi, J.; Liu, Y.; et al. Tuning of optical behavior in monolayer and bilayer molybdenum disulfide using hydrostatic pressure. *J. Phys. Chem. Lett.* **2022**, *13*, 161–167. [CrossRef]

29. Qiao, W.; Sun, H.; Fan, X.; Jin, M.; Liu, H.; Tang, T.; Xiong, L.; Niu, B.; Li, X.; Wang, G. Interlayer coupling and pressure engineering in bilayer MoS_2. *Crystals* **2022**, *12*, 693. [CrossRef]
30. Michail, A.; Parthenios, J.N.; Anestopoulos, D.; Galiotis, C.; Christian, M.; Ortolani, L.; Morandi, V.; Papagelis, K. Controllable, eco–friendly, synthesis of highly crystalline 2D–MoS_2 and clarification of the role of growth–induced strain. *2D Mater.* **2018**, *5*, 035035. [CrossRef]
31. Mao, H.K.; Xu, J.; Bell, P.M. Calibration of the ruby pressure gauge to 800 kbar under quasi–hydrostatic conditions. *J. Geophys. Res.* **1986**, *91*, 4673–4676. [CrossRef]
32. Mujica, A.; Rubio, A.; Muñoz, A.; Needs, R.J. High–pressure phases of group–IV, III–V, and II–VI compounds. *Rev. Mod. Phys.* **2003**, *75*, 863–912. [CrossRef]
33. Francisco-López, A.; Han, B.; Lagarde, D.; Marie, X.; Urbaszek, B.; Robert, C.; Goñi, A.R. On the impact of the stress situation on the optical properties of WSe_2 monolayers under high pressure. *Pap. Phys.* **2019**, *11*, 110005. [CrossRef]
34. Bronsema, K.D.; de Boer, J.L.; Jellinek, F. On the structure of molybdenum diselenide and disulfide. *Z. Anorg. Allg. Chem.* **1986**, *540*, 15–17. [CrossRef]
35. Momma, K.; Izumi, F. VESTA 3 for three-dimensional visualization of crystal, volumetric and morphology data. *J. Appl. Crystallogr.* **2011**, *44*, 1272–1276. [CrossRef]
36. Murata, K.; Yokogawa, K.; Yoshino, H.; Klotz, S.; Munsch, P.; Irizawa, A.; Nishiyama, M.; Iizuka, K.; Nanba, T.; Okada, T.; et al. Pressure transmitting medium Daphne 7474 solidifying at 3.7 GPa at room temperature. *Rev. Sci. Instrum.* **2008**, *79*, 085101. [CrossRef] [PubMed]
37. Sasaki, S.; Kato, S.; Kume, T.; Shimizu, H.; Okada, T.; Aoyama, S.; Kusuyama, F.; Murata, K. Elastic properties of new–pressure transmitting medium Daphne 7474 under high pressure. *Jpn. J. Appl. Phys.* **2010**, *49*, 106702. [CrossRef]
38. Piermarini, G.J.; Block, S.; Barnett, J.S. Hydrostatic limits in liquids and solids to 100 kbar. *J. Appl. Phys.* **1973**, *44*, 5377–5382. [CrossRef]
39. Jayaraman, A. Diamond anvil cell and high–pressure physical investigations. *Rev. Mod. Phys.* **1983**, *55*, 65–108. [CrossRef]
40. Molina-Sánchez, A.; Hummer, K.; Wirtz, L. Vibrational and optical properties of MoS_2: From monolayer to bulk. *Surf. Sci. Rep.* **2015**, *70*, 554–586. [CrossRef]
41. Zhang, X.; Qiao, X.-F.; Shi, W.; Wu, J.-B.; Jiang, D.-S.; Tan, P.-H. Phonon and Raman scattering of two–dimensional transition metal dichalcogenides from monolayer, multilayer to bulk material. *Chem. Soc. Rev.* **2015**, *44*, 2757–2785. [CrossRef]
42. Livneh, T.; Spanier, J.E. A comprehensive multiphonon spectral analysis in MoS_2. *2D Mater.* **2015**, *2*, 035003. [CrossRef]
43. Li, H.; Zhang, Q.; Yap, C.C.R.; Tay, B.K.; Edwin, T.H.T.; Olivier, A.; Baillargeat, D. From Bulk to Monolayer MoS_2: Evolution of Raman Scattering. *Adv. Funct. Mater.* **2012**, *22*, 1385–1390. [CrossRef]
44. Chakraborty, B.; Matte, H.S.S.R.; Sood, A.K.; Rao, C.N.R. Layer–dependent resonant Raman scattering of a few layer MoS_2. *J. Raman Spectrosc.* **2013**, *44*, 92–96. [CrossRef]
45. Aguiar Sousa, J.H.; Araújo, B.S.; Ferreira, R.S.; San–Miguel, A.; Alencar, R.S.; Filho, A.G.S. Pressure tuning resonance Raman scattering in monolayer, trilayer, and many–layer molybdenum disulfide. *ACS Appl. Nano Mater.* **2022**, *5*, 14464–14469. [CrossRef]
46. Carvalho, B.R.; Pimenta, M.A. Resonance Raman spectroscopy in semiconducting transition–metal dichalcogenides: Basic properties and perspectives. *2D Mater.* **2020**, *7*, 042001. [CrossRef]
47. Livneh, T.; Eran Sterer, E. Resonant Raman scattering at exciton states tuned by pressure and temperature in 2*H*–MoS_2. *Phys. Rev. B* **2010**, *81*, 195209. [CrossRef]
48. Jiang, J.-J.; Li, H.-P.; Dai, L.-D.; Hu, H.-Y.; Zhao, C.-S. Raman scattering of 2*H*–MoS_2 at simultaneous high temperature and high pressure (up to 600 K and 18.5 GPa). *AIP Adv.* **2016**, *6*, 035214. [CrossRef]
49. Shen, P.; Li, Q.; Zhang, H.; Liu, R.; Liu, B.; Yang, X.; Dong, Q.; Cui, T.; Liu, B. Raman and IR spectroscopic characterization of molybdenum disulfide under quasi–hydrostatic and non–hydrostatic conditions. *Phys. Status Solidi B* **2017**, *254*, 1600798. [CrossRef]
50. Tang, C.; Liang, L.; Zhu, X.; Liu, W.; Yang, Q.; Zhou, X.; Yan, L.; Tan, W.; Lu, M.; Lu, M. Theoretical and experimental Raman study of molybdenum disulfide. *J. Phys. Chem. Solids* **2021**, *156*, 110154. [CrossRef]
51. Aksoy, R.; Ma, Y.; Selvi, E.; Chyu, M.C.; Ertas, A.; White, A. X–ray diffraction study of molybdenum disulfide to 38.8 GPa. *J. Phys. Chem. Solids* **2006**, *67*, 1914–1917. [CrossRef]
52. Bandaru, N.; Kumar, R.S.; Sneed, D.; Tschauner, O.; Baker, J.; Antonio, D.; Luo, S.-N.; Hartmann, T.; Zhao, Y.; Venkat, R. Effect of pressure and temperature on structural stability of MoS_2. *J. Phys. Chem. C* **2014**, *118*, 3230–3235. [CrossRef]
53. Bagnall, A.G.; Liang, W.Y.; Marseglia, E.A.; Welber, B. Raman studies of MoS_2 at high pressure. *Phys. B+C* **1980**, *99*, 343–346. [CrossRef]
54. Sugai, S.; Ueda, T. High–pressure Raman spectroscopy in the layered materials 2*H*–MoS_2, 2*H*–$MoSe_2$, and 2*H*–$MoTe_2$. *Phys. Rev. B* **1982**, *26*, 6554–6558. [CrossRef]
55. Pimenta, M.A.; del Corro, E.; Carvalho, B.R.; Fantini, C.; Malard, L.M. Comparative study of Raman spectroscopy in graphene and MoS_2-type transition metal dichalcogenides. *Acc. Chem. Res.* **2015**, *48*, 41–47. [CrossRef]

56. Lee, C.; Yan, H.; Brus, L.E.; Heinz, T.F.; Hone, J.; Ryu, S. Anomalous lattice vibrations of single- and few-layer MoS_2. *ACS Nano* **2010**, *4*, 2695–2700. [CrossRef]
57. Luo, X.; Zhao, Y.; Zhang, J.; Xiong, Q.; Quek, S.Y. Anomalous frequency trends in MoS_2 thin films attributed to surface effects. *Phys. Rev. B* **2013**, *88*, 075320. [CrossRef]
58. Scheuschner, N.; Ochedowski, O.; Kaulitz, A.-M.; Gillen, R.; Schleberger, M.; Maultzsch, J. Photoluminescence of freestanding single– and few–layer MoS_2. *Phys. Rev. B* **2014**, *89*, 125406. [CrossRef]
59. Wang, Y.; Cong, C.; Qiu, C.; Yu, T. Raman spectroscopy study of lattice vibration and crystallographic orientation of monolayer MoS_2 under uniaxial strain. *Small* **2013**, *9*, 2857–2861. [CrossRef]
60. Rodríguez, Á.; Çakıroğlu, O.; Li, H.; Carrascoso, F.; Mompean, F.; Garcia-Hernandez, M.; Munuera, C.; Castellanos-Gomez, A. Improved strain transfer efficiency in large-area two-dimensional MoS_2 obtained by gold-assisted exfoliation. *J. Phys. Chem. Lett.* **2024**, *15*, 6355–6362. [CrossRef]
61. He, X.; Montakim Tareq, A.; Qi, K.; Conti, Y.; Tung, V.; Chiang, N. High-resolution distance dependence interrogation of scanning ion conductance microscopic tip-enhanced Raman spectroscopy enabled by two-dimensional molybdenum disulfide substrates. *Nano Lett.* **2024**, *24*, 13805–13810. [CrossRef] [PubMed]
62. Filintoglou, K.; Papadopoulos, N.; Arvanitidis, J.; Christofilos, D.; Frank, O.; Kalbac, M.; Parthenios, J.; Kalosakas, G.; Galiotis, C.; Papagelis, K. Raman spectroscopy of graphene at high pressure: Effects of the substrate and the pressure transmitting media. *Phys. Rev. B* **2013**, *88*, 045418. [CrossRef]
63. Ves, S.; Cardona, M. A new application of the diamond anvil cell: Measurements under uniaxial stress. *Solid State Commun.* **1981**, *38*, 1109–1112. [CrossRef]
64. Blacha, A.; Ves, S.; Cardona, M. Effects of uniaxial strain on the exciton spectra of CuCl, CuBr, and CuI. *Phys. Rev. B* **1983**, *27*, 6346–6362. [CrossRef]
65. Nye, J.F. *Physical Properties of Crystals: Their Representation by Tensors and Matrices*; Oxford University Press: Oxford, UK, 1985.
66. Anastassakis, E.; Cardona, M. Phonons, strains, and pressure in semiconductors. *Semicond. Semimet.* **1998**, *55*, 117–233. [CrossRef]
67. Feldman, J. Elastic constants of $2H$–MoS_2 and $2H$–$NbSe_2$ extracted from measured dispersion curves and linear compressibilities. *J. Phys. Chem. Solids* **1976**, *37*, 1141–1144. [CrossRef]
68. Pierre-Louis, O. Adhesion of membranes and filaments on rippled surfaces. *Phys. Rev. E* **2008**, *78*, 021603. [CrossRef] [PubMed]
69. Nicolle, J.; Machon, D.; Poncharal, P.; Pierre-Louis, O.; San-Miguel, A. Pressure–mediated doping in graphene. *Nano Lett.* **2011**, *11*, 3564–3568. [CrossRef] [PubMed]
70. Webb, A.W.; Feldman, J.L.; Skelton, E.F.; Towle, L.C.; Liu, C.Y.; Spain, I.L. High pressure investigations of MoS_2. *J. Phys. Chem. Solids* **1976**, *37*, 329–335. [CrossRef]
71. Conley, H.; Wang, B.; Ziegler, J.; Haglund, R.F.; Pantelides, S.T.; Bolotin, K.I. Bandgap engineering of strained monolayer and bilayer MoS_2. *Nano Lett.* **2013**, *13*, 3626–3630. [CrossRef] [PubMed]
72. Fan, X.; Chang, C.-H.; Zheng, W.; Kuo, J.-L.; Singh, D.J. The electronic properties of single–layer and multilayer MoS_2 under high pressure. *J. Phys. Chem. C* **2015**, *119*, 10189–10196. [CrossRef]
73. Zhao, L.; Jiang, Y.; Li, C.; Liang, Y.; Wei, Z.; Wei, X.; Zhang, Q. Probing anisotropic deformation and near–infrared emission tuning in thin–layered InSe crystal under high pressure. *Nano Lett.* **2023**, *23*, 3493–3500. [CrossRef]
74. Michail, A.; Delikoukos, N.; Parthenios, J.; Galiotis, C.; Papagelis, K. Optical detection of strain and doping inhomogeneities in single layer MoS_2. *Appl. Phys. Lett.* **2016**, *108*, 173102. [CrossRef]
75. Murata, K.; Goto, K.; Klotz, S.; Bhoi, D.; Béneut, K.; Uwatoko, Y.; Yoshino, H.; Aoki, S. A new pressure medium, Daphne 7676, with solidification pressure of 5 GPa at 300 K and low melting point of around 110 K at ambient pressure. *High Press. Res.* **2025**, *45*, 151–161. [CrossRef]
76. Proctor, J.E.; Gregoryanz, E.; Novoselov, K.S.; Lotya, M.; Coleman, J.N.; Halsall, M.P. High–pressure Raman spectroscopy of graphene. *Phys. Rev. B* **2009**, *80*, 073408. [CrossRef]

Disclaimer/Publisher's Note: The statements, opinions and data contained in all publications are solely those of the individual author(s) and contributor(s) and not of MDPI and/or the editor(s). MDPI and/or the editor(s) disclaim responsibility for any injury to people or property resulting from any ideas, methods, instructions or products referred to in the content.

crystals

MDPI

Article

A Novel High-Performance 2-to-4 Decoder Design Utilizing a Plasmonic Well and Suspended Graphene Nanoribbon

Mohammad Javad Maleki * and Mohammad Soroosh *

Department of Electrical Engineering, Shahid Chamran University of Ahvaz, Ahvaz 6135783151, Iran
* Correspondence: mj.maleki@scu.ac.ir (M.J.M.); m.soroosh@scu.ac.ir (M.S.); Tel.: +98-613-322-6605 (M.S.)

Abstract: This paper presents a compact and high-performance 2-to-4 optical decoder based on a plasmonic well structure incorporating suspended graphene nanoribbons. By exploiting the tunable graphene's chemical potential, the propagation and confinement of surface plasmon polaritons are dynamically controlled, enabling efficient routing of optical signals toward the output ports. Finite-difference time-domain simulations are employed to analyze the influence of channel geometry and graphene chemical potential on surface plasmon polariton propagation, refractive index, and transmission loss. The designed structure, featuring a 30 nm wide and 10 nm high plasmonic well, achieves a low propagation loss of 0.188 dB/μm and a high figure of merit of 1950 at 40 THz. The designed decoder exhibits a contrast ratio of 36.93 dB and crosstalk suppression of −36.93 dB, while occupying a remarkably small area (0.05 μm^2), demonstrating superior optical performance and compactness compared to previously reported designs. These results confirm the potential of the proposed plasmonic well-based decoder as a fundamental component for next-generation nanoscale optical and plasmonic computing systems.

Keywords: optical decoder; plasmonic well; surface plasmon polariton; suspended graphene

1. Introduction

High-speed processing, commonly referred to as high-performance computing, has become the backbone of today's digital world and an indispensable technological requirement. The growing demand for rapid computation and data analysis arises from the explosion of big data and the unprecedented rate of information generation, where processing speed is essential for extracting meaningful insights. In particular, artificial intelligence and machine learning, especially the training of deep and complex models, require enormous computational power to process massive datasets and achieve high accuracy. Similarly, complex simulations and modeling tasks across diverse scientific and engineering domains, including medicine, meteorology, and industrial design, depend heavily on advanced processing capabilities. Real-time applications such as financial fraud detection and autonomous systems also rely on instantaneous data handling. High processing speed not only enhances efficiency and productivity but also accelerates innovation, boosts competitiveness, and meets the expectations of modern users accustomed to immediacy and convenience.

Optical processors are emerging as a highly promising solution to these challenges. By exploiting the speed of light (photons) instead of electrons for information transmission and processing, optical systems offer transformative potential for future computing technologies. Their major advantages include ultra-high signal propagation speed, minimal delay, low energy consumption due to reduced thermal losses, massive parallelism ideal for matrix-based computations in AI, and reduced signal interference, all of which contribute to superior computational performance [1,2].

Surface plasmon polaritons (SPPs) represent a fascinating optoelectronic phenomenon in which terahertz waves transmit along the metal (graphene)-dielectric interface. These waves exhibit hybrid characteristics of light and matter, coupling photons with the collective oscillations of free electrons at the metal surface [3]. One of their most remarkable properties is their ability to confine electromagnetic fields to dimensions far below the diffraction limit, enabling manipulation of light at the nanoscale [4]. Owing to these characteristics, SPPs have found extensive application in nanophotonic circuits, ultrasensitive biosensors, and high-density data storage, as they enable high-speed signal transmission and processing in extremely compact geometries [5–7].

Graphene has emerged as a superior material for plasmonic devices, outperforming traditional metals such as gold and silver. Its exceptional electronic and optical properties, combined with its two-dimensional atomic structure, provide low optical loss and high plasmon tunability in the infrared to terahertz range. Most importantly, graphene's charge carrier concentration and thus its optical properties, such as refractive index and conductivity, can be dynamically controlled through applied gate voltage [8]. This tunability allows for the realization of active plasmonic components, including modulators [9,10], switches [11,12], and sensors [13,14], which are not feasible with conventional metals. Furthermore, graphene's atomic thinness enables stronger field confinement and enhanced light–matter interaction at the nanoscale, offering tremendous potential for miniaturized and highly integrated optical systems. These unique characteristics make graphene an ideal candidate for next-generation plasmonic and electro-optic technologies.

Optical decoders represent a new class of devices that employ photons or SPPs, rather than electrons, for data transmission and logic operations. Similar to electronic decoders, they convert binary input codes into single active outputs but do so with markedly greater speed and energy efficiency [15]. In SPP-based decoders, electrical input signals modulate the optical properties of materials such as refractive index, thereby steering light propagation within waveguides to the desired output port. Because SPP propagation is governed by the speed of light, these devices achieve far higher performance than their electronic counterparts, which are limited by carrier mobility, bandwidth constraints, and electrical interference. Such characteristics make SPP-based decoders particularly suitable for emerging applications in ultrafast communication networks and advanced optical computing [16,17].

To date, only a few SPP-based decoders have been introduced. Maleki et al. [18] proposed a 1-to-2 decoder utilizing a graphene nanoribbon on a silicon dioxide substrate, achieving a contrast ratio of 45.73 dB, a coupling length of 1.99 μm, and a device footprint of 1.92 μm^2 with a figure of merit (FOM) of 975. Later, Maleki et al. [19] designed a 2-to-4 decoder that exploits the tunable chemical potential of graphene to manipulate the SPP wave vector within a graphene-silicon plasmonic channel. This design demonstrated a crosstalk level of −36.65 dB and a contrast ratio of 36.43 dB within a compact 2.1 μm^2 footprint. More recently, Maleki et al. [20] reported a 2-to-4 plasmonic decoder using a 60 nm wide graphene nanoribbon and a silicon dioxide channel optimized for minimal loss (0.268 dB/μm). The device successfully decoded optical information from two inputs to four outputs, exhibiting a 37.11 dB contrast ratio and −37.13 dB crosstalk while maintaining an ultra-compact area of 1.08 μm^2.

Han et al. [21] use a graphene-integrated metasurface that implements dual-peak electromagnetically induced transparency. The device operation is based on resonant interactions (bright/dark resonators) that create two sharp transmission peaks (dual-peak EIT) in the spectral domain; logical decoding exploits the presence/absence or shifting of those resonance peaks (frequency-selective transmission) by electrically tuning the graphene. Metasurface relies on localized resonances of patterned metasurface elements (sub-wavelength resonators) that couple to incident THz waves. Coupling between a "bright" resonator (which couples strongly to free-space excitation) and a "dark" resonator (which couples weakly) produces EIT/PIT. Tuning changes the resonant frequency and/or the coupling strength of resonators (via graphene conductivity). The output logic depends on whether the input frequency lies within a transparency peak (transmits to a detector/port) or is blocked, so the device is inherently frequency-domain and typically narrowband (resonant peaks). Dual-peak EIT means two distinct transparency resonances can be engineered; decoding is achieved by shifting or enabling/disabling these peaks via tuning graphene's Fermi level (which changes resonance conditions). Operation is therefore spectral, and frequency channels carry the logic information.

The innovative design presented in this study is a high-performance 2-to-4 optical decoder that utilizes a plasmonic well structure incorporating suspended graphene nanoribbons. This novel configuration is specifically engineered to overcome the limitations of previous plasmonic decoders by enabling superior light confinement and active control over the Surface Plasmon Polariton wave propagation through the tunable chemical potential of the graphene. The proposed design demonstrates performance metrics that confirm its superiority over previous similar research, achieving a minimal propagation loss of just 0.188 dB/μm and an exceptionally high FOM of 1950 at a frequency of 40 THz, substantially exceeding that reported in earlier designs [18–20]. The device also exhibits a high contrast ratio of 36.93 dB and excellent crosstalk suppression of −36.93 dB, all while maintaining an ultra-compact footprint, confirming its viability as a fundamental component for integrated optoelectronic systems. The following sections will detail these results: Section 2 introduces the fundamental structure and the physics of the well-based plasmonic channel; Section 3 provides the comprehensive simulation-based results and analysis of the design; Section 4 presents the full architecture and operational verification of the proposed 2-to-4 plasmonic decoder; and Section 5 summarizes the conclusions and impact of this study.

2. The Well-Based Plasmonic Structure

Figure 1a shows a fundamental structure for transmission and directing surface plasmon polaritons from G1 nanoribbon to G2 and G3 nanoribbons. G2 and G3 direct the surface plasmon polaritons toward O0 and O1 output ports. The bottom layer is a 200 nm silicon substrate, followed by a 40 nm gold deposition, a 100 nm silicon dioxide layer, and finally, three graphene nanoribbons. The applied voltage to these nanoribbons controls the SPP transmission via setting the Fermi level energy. It can be seen that Figure 1a consists of three channels, including the plasmonic wells, where the SPPs excited in the G1 can penetrate into the G2 and G3, which are 10 nm apart, due to the transverse component of the electric field, and are confined near the graphene surface based on the fermi adjustment and propagate in the z-direction. A plasmonic well is considered for guiding and confining the SPPs, as shown in Figure 1b. This figure shows a two-dimensional view of a channel (including a plasmonic well) in the xy-plane.

Surface plasmon polaritons are excited when transverse magnetic (TM)-polarized light illuminates upon the suspended graphene. As described in Section 2, the TM illumination provides an electric field component normal to the graphene surface and a magnetic field component along the x-direction, satisfying the boundary conditions required for SPP excitation. The interaction between the incident TM field and the collective oscillations of free charge carriers in graphene induces surface current densities, governed by the graphene's complex surface conductivity modeled using the Kubo formalism. This coupling results in the formation of bound electromagnetic surface modes, the SPPs, that propagate along the z-direction within the plasmonic well region, i.e., the air gap between the suspended graphene layer and the gold substrate. Furthermore, by applying appropriate bias voltages to the graphene ribbons, the Fermi level (or chemical potential, μ_c) is dynamically adjusted, allowing active control over the excitation and propagation of SPPs. At higher μ_c (e.g., 0.6 eV), the real part of the effective refractive index increases and losses decrease, facilitating strong SPP excitation and efficient transmission. Conversely, at lower μ_c (0.2 eV), the field confinement weakens and losses rise sharply, suppressing SPP propagation.

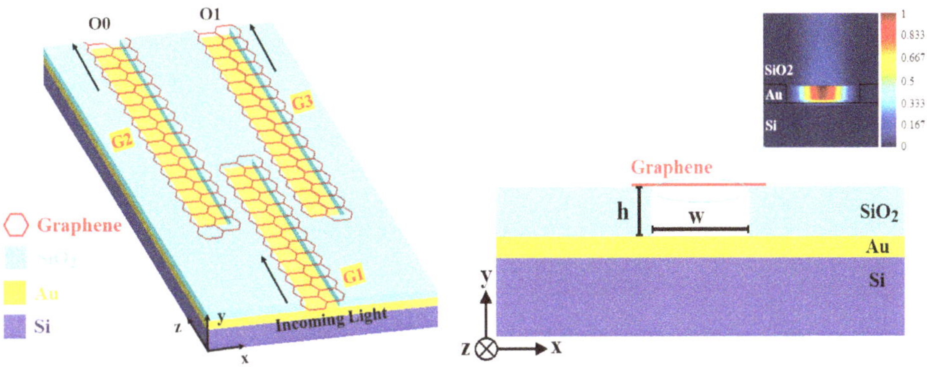

(a) (b)

Figure 1. (**a**) The proposed structure including three plasmonic wells with graphene nanoribbons for directing SPPs along the z-direction toward the output ports; (**b**) the plasmonic well with width w and height h in the xy-plane. The SPP confinement has been shown in the plasmonic channel.

When transverse magnetic light illuminates the structure, it excites SPPs, which then propagate along the z-axis in a guiding medium located beneath the graphene nanoribbon. The magnetic field within this channel (air between graphene and gold) can be mathematically expressed.

$$H_x = Ae^{-ik_1y} + Be^{ik_1y} \tag{1}$$

The top air and silicon, the magnetic component is expressed as below:

$$H_x = \begin{cases} Ce^{-ik_2y} & \text{for above the channel} \\ De^{ik_3(h+y)} & \text{for silicon} \end{cases} \tag{2}$$

where k denotes the wavenumber, and the subscripts 1, 2, and 3 correspond to the channel, the region above the channel, and the silicon layer, respectively. The coefficients A through D represent the amplitudes of the guided modes. By applying above equations, the wave equations can be reduced to the form given in Equation (3) [22].

$$H_x = \left(-j\omega\varepsilon_0\varepsilon_i k_i^{-1}\right)E_y,\ \frac{\partial}{\partial y}H_x = (-j\omega\varepsilon_0\varepsilon_i)E_z \tag{3}$$

Within the channel, the electric fields E_y and E_z corresponding to the y and z directions are obtained from Equation (4). In this expression, ω denotes the angular frequency, ε is the permittivity, and the subscripts 0 and i (with i spanning 1 to 3) correspond to free space and the previously defined regions, respectively.

$$E_y = \left(\frac{jAk_1}{\omega\varepsilon_1\varepsilon_0}\right)e^{-jk_1y} + \left(\frac{-jBk_1}{\omega\varepsilon_1\varepsilon_0}\right)e^{jk_1(y+h)} \tag{4a}$$

$$E_z = \left(\frac{Ak_1}{\omega\varepsilon_1\varepsilon_0}\right)e^{-jk_1y} + \left(\frac{-Bk_1}{\omega\varepsilon_1\varepsilon_0}\right)e^{jk_1(y+h)} \quad (4b)$$

At the interface of y = 0, the boundary conditions are defined as $E_z^1 = E_z^2$ and $H_x^1 = H_x^2 + \sigma_g E_z$. Similarly, at the interface of y = −h, the boundary conditions are $E_z^1 = E_z^3$ and $H_x^1 = H_x^3$. The presence of the graphene ribbon is crucial in determining these boundary conditions due to its complex conductivity, represented by σ_g. The conductivity of graphene is modeled using Kubo's approach, which considers both the interband and intraband contributions. The component of graphene's permittivity parallel to the surface is given by Equation (5) [23].

$$\varepsilon_g^{\|}(\omega,\mu_c) = \varepsilon_r + i\sigma_g(\omega,\mu_c)(\omega\varepsilon_0\gamma)^{-1} \quad (5)$$

Both the graphene thickness (γ) and its chemical potential (μ_c) are key factors determining the characteristics of the optical fields. Calculations are performed using the 3D finite-difference time-domain (FDTD) method, employing a 0.1 nm mesh size and a 5.6 attosecond simulation step to maintain accuracy. These amounts satisfy the Courant condition for achieving the convergence.

3. Results and Discussion

Manipulating the graphene nanoribbon's chemical potential enables control over SPP transmission within the channel, which lies between the nanoribbon and a plasmonic well. Figure 2a demonstrates how channel transmission loss changes with varying chemical potentials (0.2 eV, 0.4 eV, and 0.6 eV) across different frequencies, ranging from 58.73 to 0.28 dB/μm. A significant finding is that increasing the graphene chemical potential leads to decreased transmission loss. This effect arises from the direct impact of the chemical potential on the channel's effective refractive index. Figure 2b visually presents the real ($Re(n_{eff})$) and imaginary ($Im(n_{eff})$) parts of the effective refractive index for various chemical potentials. This figure shows that for chemical potentials between 0.2 eV and 0.6 eV at 40 THz, the real part varies from 191.73 to 45.11 and the imaginary part from 54.8 to 0.0136.

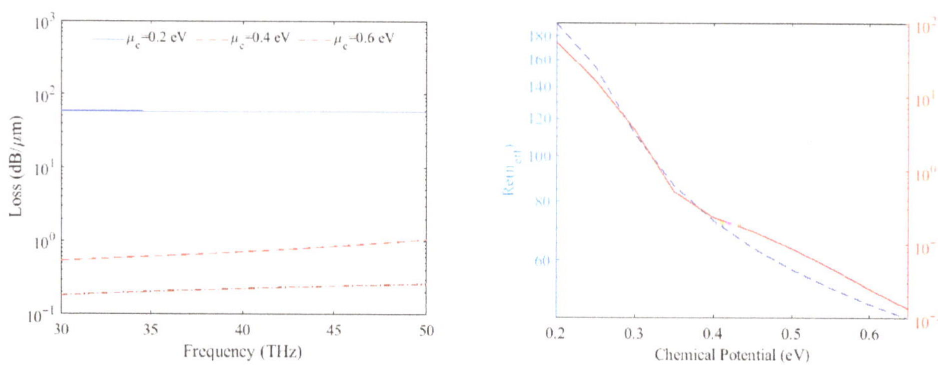

(a) (b)

Figure 2. Visualizing key performance metrics: (**a**) Transmission losses versus frequency are plotted for distinct graphene chemical potentials. (**b**) The figure reveals the relationship between graphene chemical potential and the channel's effective refractive index, detailing its imaginary and real components.

The degree of confinement in the channel is directly determined by the magnitude of the refractive index's real part. An increased real component promotes stronger localization of surface plasmon polaritons in the channel and suppresses field spreading in the x-direction. Meanwhile, the imaginary component represents transmission loss, with higher values leading to more pronounced attenuation. This relationship is formalized by Equation (6) for transmission loss (α) [24]:

$$\alpha = \left(\frac{2\omega}{c}\right)\text{Im}(n_{eff}) \tag{6}$$

The parameter 'c' refers to the speed of light in a vacuum. The chemical potential of the graphene layer can be precisely controlled by varying the electrical potential difference applied between the metal and the suspended graphene. By adjusting this bias voltage, the carrier concentration within the graphene changes, leading to modifications in its Fermi level and, consequently, its optical and electronic response. This tunability enables dynamic regulation of the device's performance parameters, such as light absorption and modulation efficiency, making it particularly advantageous for high-speed and reconfigurable optoelectronic applications. A capacitor-based theoretical framework, presented in Equation (7), elucidates the correlation between the graphene's chemical potential and the applied voltage (V_b) [25].

$$\mu_c(V) = \hbar v_F \left[\frac{2.25 \times 10^{16}(V_b - 0.8)}{\pi}\right]^{0.5} \tag{7}$$

This analysis incorporates Planck's constant and the Fermi velocity as fundamental coefficients [25]. As depicted in Figure 2a, the ability of terahertz waves to transmit through the plasmonic channel is directly dependent on the graphene's chemical potential (μ_c). When μ_c is 0.6 eV, SPP transmission is facilitated. Conversely, at μ_c = 0.2 eV, substantial losses prevent plasmon propagation beneath the graphene nanoribbons. In this study, a graphene chemical potential of 0.6 eV is selected to facilitate SPP transmission, while a reduced value of 0.2 eV is employed to effectively suppress their propagation.

According to Equation (7), graphene's chemical potential can be precisely controlled by applying bias voltages. In particular, applying bias voltages of 1.202 V and 4.418 V results in chemical potentials of μ_c = 0.2 eV and μ_c = 0.6 eV in the graphene ribbons, respectively. For clarity, the chemical potentials associated with the three graphene nanoribbons (G1, G2, and G3 in Figure 1a) are denoted as μ_{c1}, μ_{c2}, and μ_{c3}.

Figure 3 demonstrates the directional control of SPPs as they transmit through waveguide G1, where its chemical potential μ_{c1} is maintained at 0.6 eV. The SPPs' destination, either output port O0 or O1, is determined by the chemical potentials of adjacent waveguides:

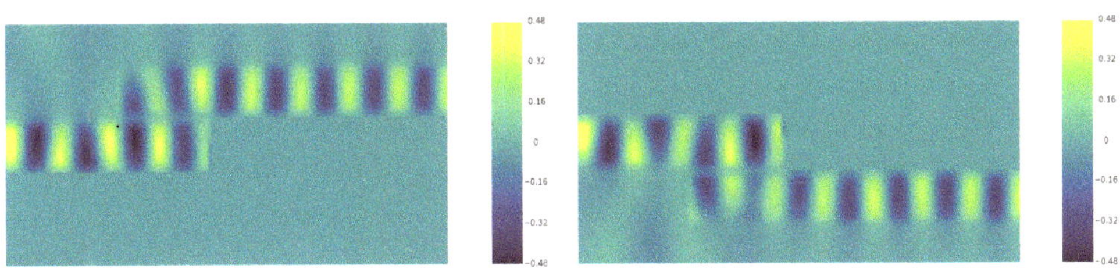

(**a**) (**b**)

Figure 3. The electric field patterns of SPP waves in the xz-plane, highlighting the impact of varying chemical potentials. Section (**a**) shows the distribution when the chemical potentials are set to μ_{c2} = 0.6 eV and μ_{c3} = 0.2 eV; while section (**b**) displays the field distribution for μ_{c2} = 0.2 eV and μ_{c3} = 0.6 eV.

(a) Activating Output O0: When μ_{c2} is 0.6 eV and μ_{c3} is 0.2 eV, SPPs successfully couple from G1 into waveguide G2, activating the output port O0, as illustrated in Figure 3a.

(b) Activating Output O1: If μ_{c2} is set to 0.2 eV and μ_{c3} to 0.6 eV, the SPPs arriving at the end of G1 are routed into waveguide G3, subsequently reaching and illuminating output port O1. This scenario is depicted in Figure 3b, which, along with Figure 3a, visually details the electric field distribution of the SPP waves throughout the entire structure.

A key factor in plasmon propagation is the height of the plasmonic well, as it dictates the channel's effective refractive index. Figure 4 illustrates that past a 10 nm height, further increases result in only minor changes to both $Re(n_{eff})$ and propagation loss. Consequently, we selected 10 nm as the proper channel height. Since the width is in the direction perpendicular to the electric field confinement, its variation has only a minor influence on field localization unless it becomes comparable to the SPP wavelength (λ_{SPP}). For the designed frequency of 40 THz, λ_{SPP} is typically on the order of a few hundred nanometers, much larger than the 30 nm channel width. Thus, variations in width below ~30 nm have minimal impact on confinement and effective refractive index. For narrow widths ($w <$ 50 nm), only the fundamental TM mode can exist. Thus, even substantial reductions in width will not change the number of supported modes, keeping field confinement nearly constant. Wider channels would risk supporting higher-order modes, increasing modal crosstalk and design complexity. A 30 nm wide channel ensures a subwavelength footprint, supporting ultra-dense integration in nanoscale photonic circuits (total device area 0.05 μm^2). Increasing width would enlarge the device without providing a measurable confinement advantage.

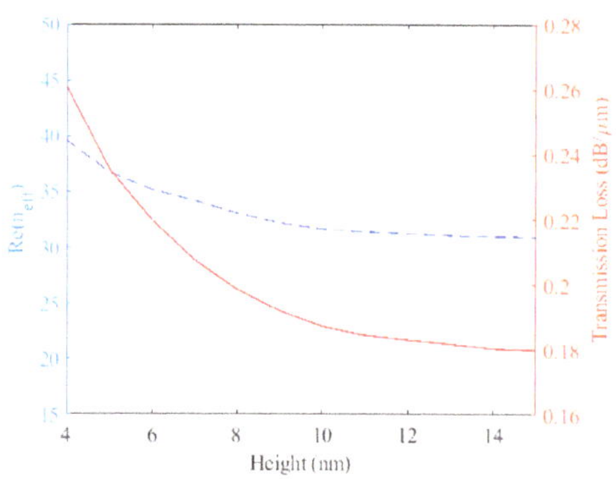

Figure 4. The loss and the real component of index in terms of the channel height. All results were taken with a constant channel width of 30 nm.

The plasmonic well was designed with a width (w) of 30 nm and a height (h) of 10 nm, based on comprehensive simulation results. These dimensions directly influence the waveguide's performance, leading to loss of 0.188 dB/μm for μ_c = 0.6 eV and 58.32 dB/μm for μ_c = 0.2 eV. To quantify the channel's efficiency in guiding surface plasmon polaritons, we use the figure of merit (FOM), which is the ratio of $Re(n_{eff})$ to $Im(n_{eff})$ [26]. A higher FOM indicates superior SPP confinement and lower propagation loss. For our specific design, operating at 40 THz, $Re(n_{eff})$ is 48.36 and $Im(n_{eff})$ is 0.0248, culminating in an FOM of 1950.

Although a high figure of merit is desirable for designing compact optical components, the inter-channel distance critically impacts the efficiency of terahertz wave coupling. Suboptimal spacing can induce detrimental coupling, leading to interference and a degradation in device functionality. To evaluate this, the coupling length, which denotes the distance over which complete optical power is transferred between neighboring channels, is calculated using Equation (8). A longer coupling length is advantageous, as it directly correlates with reduced side-field penetration and, consequently, diminished interference [27].

$$L_c = \pi(\beta_e - \beta_o)^{-1} \tag{8}$$

The z-component of the wave vector is denoted by β, with 'e' and 'o' subscripts distinguishing even and odd modes. As presented in Figure 5, the coupling length (L_c) exhibits an exponential increase as the distance between adjacent waveguides expands. This observed increase in L_c is partly attributable to the tight electric field confinement within the channel. While increasing channel separation is an effective strategy for minimizing interference, it is crucial to balance this with the need for a compact overall structure. Accordingly, a minimum inter-channel spacing of 40 nm has been chosen to prevent unwanted coupling.

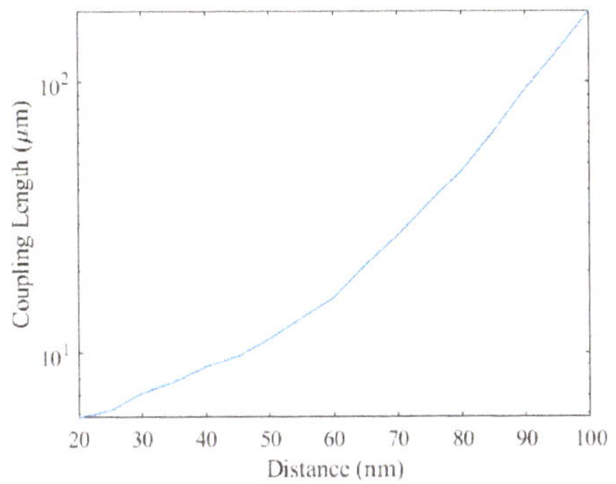

Figure 5. The coupling length for varying separation distances at a chemical potential of μ_c = 0.6 eV. In this configuration, each channel has a width of 30 nm and a height of 10 nm.

Our simulations reveal that manipulating μ_c is an effective way to control SPP propagation. For μ_c = 0.2 eV, plasmons face significant absorption and cannot pass through the channel. Conversely, setting the chemical potential to 0.6 eV enables optical waves to propagate efficiently. This allows us to guide surface plasmon polaritons toward either output O0 or O1, depending on the specific chemical potential values applied to graphene nanoribbons G1, G2, and G3. Figure 3 serves to validate this design as a foundational subwavelength switch for logic applications. Next, we will propose a 2-to-4 plasmonic decoder based on these principles.

4. The 2-to-4 Well-Based Plasmonic Decoder

This section details the design of a 2-to-4 subwavelength plasmonic decoder, which is an extension of the basic structure introduced in Section 2 (Figure 6a). The device has two inputs, U0 and U1, whose states are determined by the chemical potentials of their corresponding graphene nanoribbons. For guiding surface plasmon polaritons to the four output ports (O0 to O3), the decoder utilizes six waveguides (W1 through W6), as depicted in Figure 6b. Control over the SPP pathways within these waveguides is achieved via five graphene nanoribbons. Figure 6b specifically identifies the chemical potentials U0, U1, U0′, and U1′ with their associated nanoribbons. Consistent with the approach discussed in Section 2, only two specific chemical potential values, 0.2 eV and 0.6 eV, are used to switch and direct the SPP waves. This operation adheres to a complementary rule: when U1 (or U0) is set to 0.2 eV, the corresponding U1′ (or U0′) is set to 0.6 eV, and the reverse applies as well.

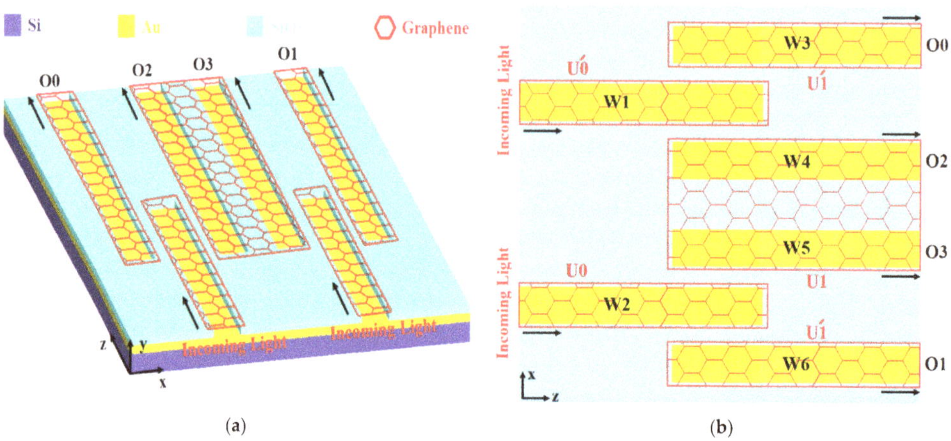

Figure 6. (**a**) Proposed decoder including a well-based channel; (**b**) top view of the structure with graphene ribbons. SPPs are directed to a specific output port based on the values of U0, U0′, U1, and U1′.

The selection of a specific output port in the decoder is determined by its truth table, which interprets the electrical inputs U0 and U1. Table 1 shows the operating conditions for this decoder. For digital signaling, a chemical potential of μ_c = 0.6 eV represents a logic 1, whereas μ_c = 0.2 eV indicates a logic 0 for the inputs.

Table 1. The working cases of the designed structure for achieving the decoding operation.

Working Case	Input Logic	Input Value	Output Logic			
		U_1 (eV), U_0 (eV)	O3	O2	O1	O0
1	[0, 0]	0.2, 0.2	0	0	0	1
2	[0, 1]	0.2, 0.6	0	0	1	0
3	[1, 0]	0.6, 0.2	0	1	0	0
4	[1, 1]	0.6, 0.6	1	0	0	0

The operation of the decoder ensures that only one output port is active at a time, depending on the input signals.

Activating O0: Setting U1 and U0 to 0.2 eV allows SPPs to move efficiently through W1 and W3, reaching output port O0, as illustrated in Figure 7a.

Activating O1: At input potentials of U1 = 0.2 eV and U0 = 0.6 eV, optical waves follow the path from W2 to W6, activating O1 while keeping all other output ports inactive (Figure 7b).

Activating O2: Output port O2 becomes active when SPPs are transmitted through waveguides W1 and W4. This occurs under the condition U1 = 0.6 eV and U0 = 0.2 eV, as shown in Figure 7c.

Activating O3: At U1 = U0 = 0.6 eV, SPPs successfully propagate through W2 and W5 to reach O3, utilizing the low-loss propagation depicted in Figure 2a. This configuration activates O3, while all other outputs remain off (Figure 7d).

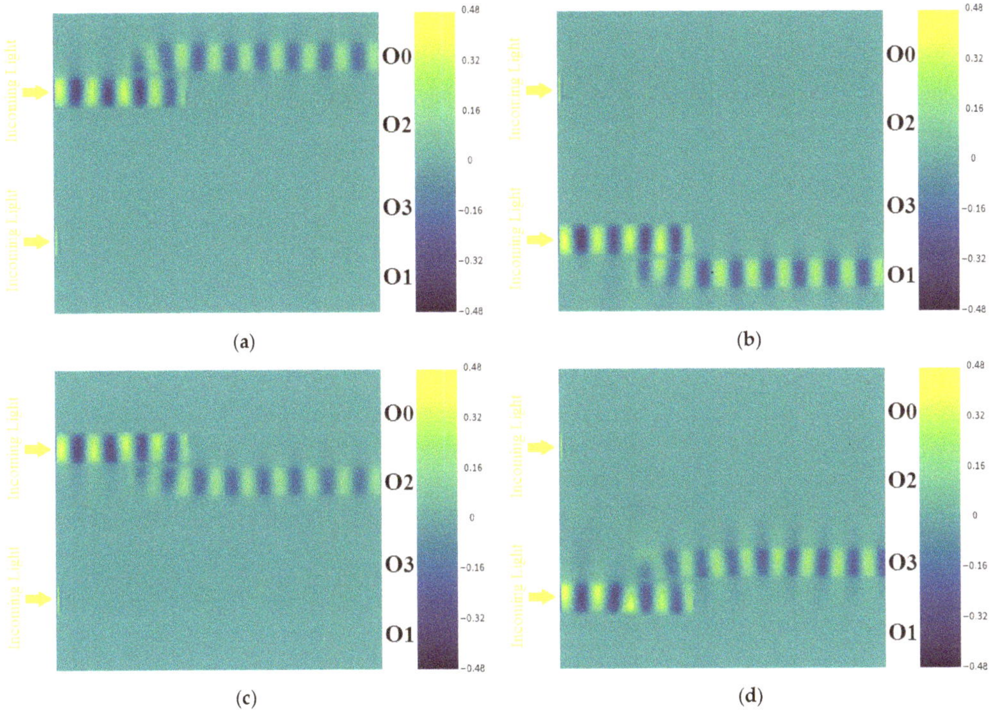

Figure 7. The field profile of SPPs for cases (U1, U0) equal to (**a**) (0.2 eV, 0.2 eV); (**b**) (0.2 eV, 0.6 eV); (**c**) (0.6 eV, 0.2 eV); and (**d**) (0.6 eV, 0.6 eV).

Table 2 provides a comprehensive look at the decoder's performance by detailing the optical power at O0 to O3 ports across four distinct operational cases. In case 1, the normalized powers for outputs O0, O1, O2, and O3 are 98.7%, 0.01%, 0.02%, and 0.01%, respectively. Case 2 shows the normalized powers of 0.01%, 98.6%, 0.01%, and 0.02% for the same output ports. For case 3, the observed normalized power values are 0.02% at O0, 0.01% at O1, 98.7% at O2, and 0.02% at O3. Finally, case 4 demonstrates power outputs of 0.01% for O0, 0.02% for O1, 0.02% for O2, and 98.7% for O3.

Table 2. Power at ports O0 to O3 corresponding to the working cases.

Working Case	Waveguide for SPP Transmission	Normalized Output Power (Logic Levels)				Margin
		O3	O2	O1	O0	
1	W1, W3	0.01% (0)	0.02% (0)	0.01% (0)	98.7% (1)	M_0
2	W2, W6	0.02% (0)	0.01% (0)	98.6% (1)	0.01% (0)	M_1, M_0
3	W1, W4	0.02% (0)	98.7% (1)	0.01% (0)	0.02% (0)	M_0
4	W2, W5	98.7% (1)	0.02% (0)	0.02% (0)	0.01% (0)	M_0

Due to the extremely compact footprint of the designed structure (approximately 0.05 μm^2), the simulated transmission spectra for the different working conditions exhibit very similar behaviors. As indicated in Table 2, the transmission efficiency for each activated output port remains consistently high, in the range of 98.6–98.7%, confirming the efficient routing capability of the plasmonic decoder. Moreover, as shown in Figure 2a, the propagation loss is minimal when the graphene chemical potential is set to μ_c = 0.6 eV, corresponding to the low-loss state of the device. To further clarify this point and strengthen the discussion, the simulated transmission results associated with Figure 7a have shown in Figure 8. These additions provide a more comprehensive understanding of the transmission performance and confirm the robustness of the proposed design across all operational conditions.

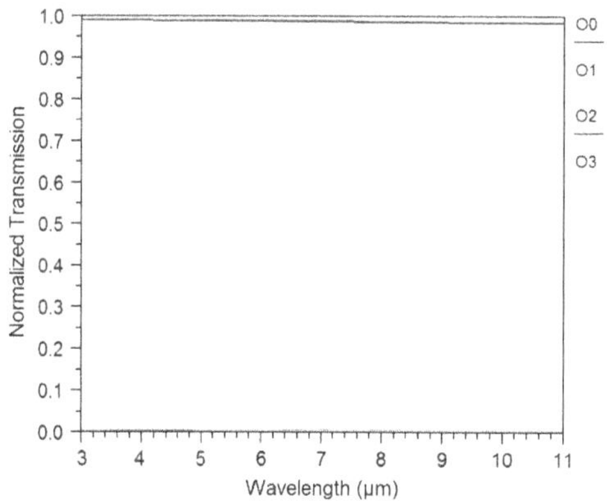

Figure 8. The normalized transmission at output ports versus the wavelength associated with Figure 7a. In this case, O0 is ON and other ports are OFF.

The logic 0 margin (M_0) and logic 1 margin (M_1) are critical parameters used to assess the designed structure's performance under the worst cases. M_0 quantifies the highest value considered as logic 0, while M_1 defines the lowest value recognized as logic 1. Based on the data in Table 2, M_0 for our structure is 0.02%, and M_1 is 98.6% (demonstrated in Case 2). To further illustrate the decoder's capability to clearly distinguish between logic 0 and logic 1, we employ the contrast ratio (CR), as detailed in Equation (9). A greater CR value directly makes more difference between the logic 0 and 1 signals at outputs [28].

$$CR = 10\log(M_1/M_0) \quad (9)$$

The proposed plasmonic device exhibits a contrast ratio of 36.93 dB, calculated based on an M_0 value of 0.02% and an M_1 value of 98.6%. Furthermore, crosstalk (CT), representing the undesirable coupling of surface plasmon polaritons between adjacent waveguides, is a significant factor in assessing overall decoder performance. Equation (10) defines crosstalk, with P_0 and P_1 representing the optical powers at the outputs for logic levels 0 and 1, respectively [29].

$$CT = 10\log(P_0/P_1) \quad (10)$$

As presented in Table 2, the crosstalk values for output ports O0 to O3 are determined to be −36.93 dB, −36.92 dB, −36.93 dB, and −36.93 dB, respectively. Consequently, the highest observed crosstalk in the structure is −36.93 dB. To effectively assess the performance of the designed decoder, it is useful to conduct a comparative analysis against other existing devices. Accordingly, Table 3 summarizes the obtained results covering the device's area, contrast ratio, crosstalk, and loss, along with corresponding data reported in previous literature.

Table 3. Comparison of the present findings with prior works.

Reference	Structure	Area (μm^2)	CR (dB)	CT (dB)	Loss (dB/μm)
[18]	Plasmonic–Graphene	1.92	45.73	-	-
[19]	Plasmonic–Graphene	2.1	36.43	−36.65	0.58
[20]	Plasmonic–Graphene	1.08	37.11	−37.13	0.268
[30]	Photonic crystal–Graphene	690	5.4	-	-
[31]	Metamaterial–Graphene	>100	23.97	-	-
[32]	Plasmonic–Teflon	0.43	12.83	-	-
This work	Plasmonic–Graphene	0.05	36.93	−36.93	0.188

The size of optical devices is a critical consideration in the design of integrated circuits. Conventional Mach-Zehnder interferometer structures are typically large on the order of several micrometers and often suffer from inefficient interference, making them unsuitable for compact optical systems [33–36]. Similarly, photonic crystal-based devices tend to be too bulky for high-density integration. Although resonant cavity-based designs can achieve some degree of miniaturization [37–40], fabricating them at sub-micrometer scales remains a significant challenge. Nevertheless, the inherent scalability of photonic crystal architectures continues to hold promise for future developments. In contrast, plasmonic structures provide a major advantage: they can process and decode optical signals at both micro and nanoscale dimensions. The proposed decoder exemplifies this capability, occupying only 0.05 μm^2, which is substantially more compact than previously reported designs [18–20,30–32]. This compactness represents one of its principal advantages. Performance-wise, the decoder also demonstrates excellent optical characteristics. Its contrast ratio of 36.93 dB indicates a strong ability to distinguish clearly between logic states 0 and 1. This CR is similar to those reported in earlier works [19,20,30–32], primarily due to the enhanced confinement of SPPs within the plasmonic well. Such tight confinement not only improves signal distinction but also results in low crosstalk, calculated at −36.93 dB.

The incorporation of suspended graphene significantly improves carrier mobility, which in turn enhances the optical confinement within the waveguide channel. This stronger confinement ensures that the electromagnetic field is more tightly bound to the propagation region, reducing energy leakage into the surrounding material. As a result, the lateral electric field experiences minimal dispersion, contributing to a remarkably low propagation loss of only 0.188 dB/μm. Such low loss is particularly advantageous for maintaining signal integrity over longer distances within compact photonic circuits. Considering the ultra-small footprint of the device, this level of loss not only supports efficient signal transmission but also allows for high-density integration of multiple optical components on a single chip. Consequently, the combination of suspended graphene and optimized channel design provides a practical and effective approach for developing low-loss, high-performance optical circuits suitable for advanced photonic applications. Overall, the proposed 2-to-4 plasmonic decoder achieves notable improvements in both compactness and transmission loss, highlighting its potential for use in advanced optical circuit technologies.

As illustrated in Figure 9, the proposed structure is created through a five-step fabrication process. It begins with a silicon substrate, followed by the deposition of a thin gold film and then a silicon dioxide layer. Grooves are precisely formed using electron beam lithography and reactive ion etching techniques [41,42]. The process concludes with the transfer of graphene ribbons [43].

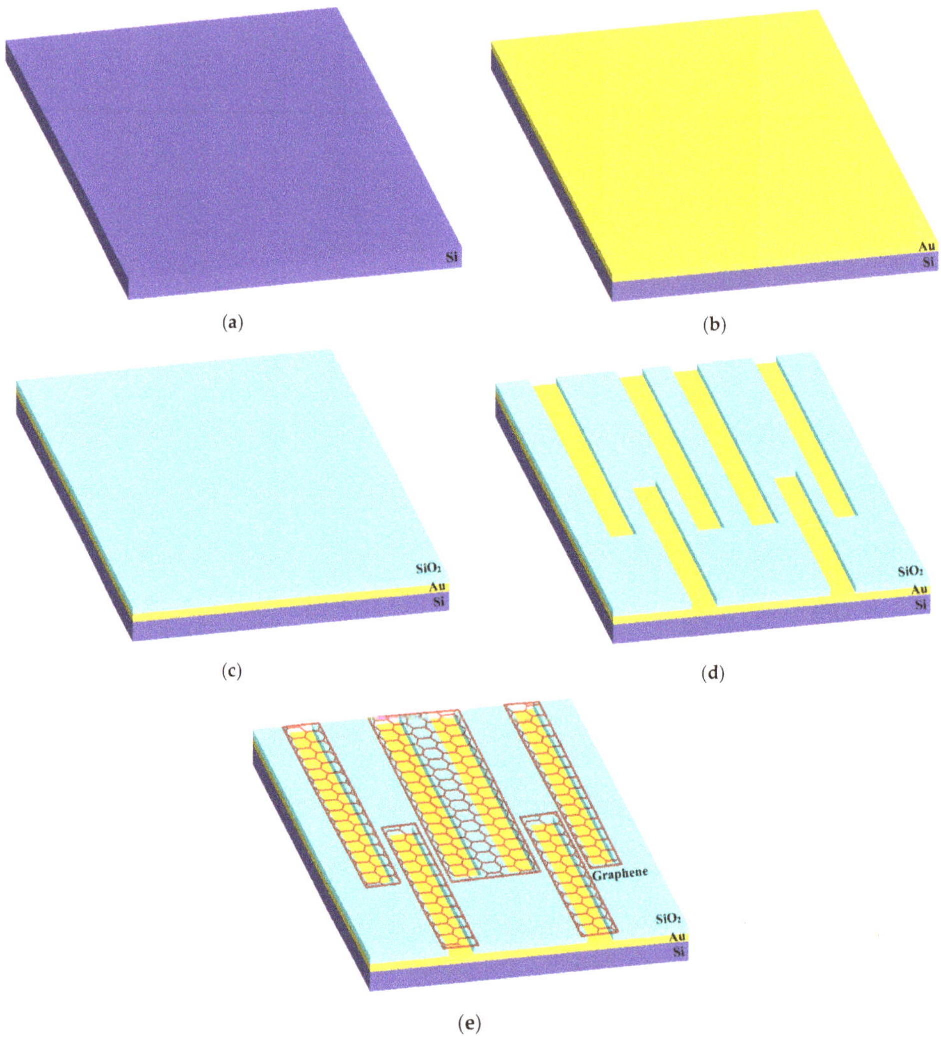

Figure 9. (**a**) The preparation of a clean silicon substrate; (**b**) a thin gold film deposition onto the substrate; (**c**) a silicon dioxide deposition; (**d**) forming grooves with etching using electron beam lithography, and the required optical pathways using reactive ion etching; (**e**) transferring graphene ribbons onto the patterned surface.

The designed 2-to-4 plasmonic decoder performs a fundamental digital logic function in the optical domain using surface plasmon polaritons. This capability is central to developing optical logic circuits, including optical multiplexers/demultiplexers, arithmetic logic units, memory addressing systems, and programmable photonic logic arrays. Because the device demonstrates high contrast ratio and low crosstalk, it can distinguish logic states with high fidelity, a critical requirement for cascaded logic operations in computing architectures.

5. Conclusions

In this study, a novel 2-to-4 plasmonic decoder utilizing suspended graphene nanoribbons and plasmonic wells was designed and analyzed. The tunability of graphene's chemical potential enabled precise control of SPP propagation and effective switching among four output ports. Simulation results demonstrated excellent performance with a high contrast ratio of 36.93 dB, low crosstalk of −36.93 dB, and minimal propagation loss of 0.188 dB/μm. The compact footprint and subwavelength operational scale make the proposed structure highly suitable for integration into future optical circuits. Compared to conventional photonic and plasmonic decoders, the device exhibits superior confinement, reduced interference, and enhanced signal distinction, primarily due to the suspended graphene configuration and optimized plasmonic well geometry. Therefore, this design offers a promising approach for realizing ultra-compact, energy-efficient, and high-speed optical logic components in advanced plasmonic and optoelectronic processors.

Author Contributions: Conceptualization, M.J.M.; methodology, M.J.M.; software, M.J.M.; validation, M.J.M., and M.S.; formal analysis, M.J.M., and M.S.; investigation, M.J.M., and M.S.; resources, M.J.M., and M.S.; data curation, M.J.M., and M.S.; writing—original draft preparation, M.J.M.; writing—review and editing, M.S.; visualization, M.J.M.; supervision, M.S.; project administration, M.S.; funding acquisition, M.S. All authors have reviewed and approved the final version of the manuscript for publication.

Funding: The authors gratefully acknowledge the financial support of Shahid Chamran University of Ahvaz through Grant No. SCU.EE1404.672.

Data Availability Statement: The original contributions presented in this study are included in the article. Further inquiries can be directed to the corresponding authors.

Conflicts of Interest: The authors confirm that they have no known competing financial or personal interests that could have influenced the work reported in this study.

Abbreviations

The abbreviations utilized throughout this manuscript are listed below:

SPPs	Surface plasmon polaritons
FOM	Figure of merit
FDTD	Finite-difference time-domain
CR	Contrast ratio
CT	Crosstalk

References

1. Weber, H.; Nakazawa, M. *Ultrahigh-Speed Optical Transmission Technology*, 1st ed.; Springer: Berlin/Heidelberg, Germany, 2007. [CrossRef]
2. Ishikawa, H. *Ultrafast All-Optical Signal Processing Devices*, 1st ed.; Wiley: Hoboken, NJ, USA, 2008.
3. Shahbazyan, T.V.; Stockman, M.I. *Plasmonics: Theory and Applications*, 1st ed.; Springer: Dordrecht, The Netherlands, 2013. [CrossRef]
4. Zhang, J.; Zhang, L.; Xu, W. Surface plasmon polaritons: Physics and applications. *J. Phys. D Appl. Phys.* **2012**, *45*, 113001. [CrossRef]
5. Liu, Y.Q.; Liu, P.K. Excitation of surface plasmon polaritons by electron beam with graphene ribbon arrays. *J. Appl. Phys.* **2017**, *121*, 113104. [CrossRef]
6. Chorsi, H.T.; Gedney, S.D. Tunable plasmonic optoelectronic devices based on graphene metasurfaces. *IEEE Photonics Technol. Lett.* **2016**, *29*, 228–230. [CrossRef]
7. Zhang, P.; Dong, Y.; Li, X.; Cao, X.; Yang, Y.; Yu, G.; Yang, S.; Wang, S.; Gong, Y. In-plane radiation of surface plasmon polaritons excited by free electrons. *Micromachines* **2024**, *15*, 723. [CrossRef]
8. Katsnelson, M.I. *The Physics of Graphene*, 2nd ed.; Cambridge University Press: Cambridge, UK, 2020. [CrossRef]
9. Kanwal, S.; Abdulsada, A.M.; Rezaei, M.H. High modulation depth, ultra-compact, and tunable electro-optical modulator based on dielectric-loaded graphene-plasmonic waveguides. *Opt. Lasers Eng.* **2025**, *193*, 109059. [CrossRef]
10. Karimkhani, H.; Karim, M.A. Wide-band high performance optical modulator based on a stack of graphene and h-BN layers with plasmonic edge mode. *Opt. Quantum Electron.* **2025**, *57*, 143. [CrossRef]

11. Maleki, M.J.; Soroosh, M.; AL-Shammri, F.K.; Alkhayer, A.G.; Mondal, H. Design and simulation of a compact subwavelength graphene-based switch for surface plasmon polariton transmission in integrated optoelectronic circuits. *Plasmonics* **2025**, *20*, 3605–3617. [CrossRef]
12. Maleki, M.J.; Soroosh, M.; Rajasekar, R. Design of a Compact All-Optical Plasmonic Switch Based on a Nonlinear Resonator for Use in the Third Telecommunication Window. *Q. J. Optoelectron.* **2024**, *7*, 55–63. [CrossRef]
13. Zhang, Z.; Lee, Y.; Haque, M.F.; Leem, J.; Hsieh, E.Y.; Nam, S. Plasmonic sensors based on graphene and graphene hybrid materials. *Nano Converg.* **2022**, *9*, 28. [CrossRef]
14. Rahad, R.; Hossain, N.; Hossain, A. Enhanced alcohol detection using surface plasmon polariton dependent MIM plasmonic sensor. *Plasmonics* **2025**, *20*, 1331–1340. [CrossRef]
15. Sharma, A.; Goswami, K.; Mondal, H.; Datta, T.; Sen, M. A review on photonic crystal based all-optical logic decoder: Linear and nonlinear perspectives. *Opt. Quantum Electron.* **2022**, *54*, 90. [CrossRef]
16. Zhu, A.; Wei, W.; Hou, W.; Cheng, L.; Hu, C. A 3–8 decoder of terahertz metamaterials and its sensing application. *Diam. Relat. Mater.* **2025**, *156*, 112411. [CrossRef]
17. Asgari, S.; Fabritius, T. Frequency-multiplexed tunable logic device based on terahertz graphene-integrated metamaterial composed of two circular ring resonator arrays. *Sci. Rep.* **2025**, *15*, 28920. [CrossRef] [PubMed]
18. Maleki, M.J.; Soroosh, M.; Akbarizadeh, G. Design and Simulation of Compact Plasmonic Decoder with High Contrast Ratio for Propagation of Graphene Surface Plasmon Polaritons. *Q. J. Optoelectron.* **2023**, *6*, 11–18. [CrossRef]
19. Maleki, M.J.; Soroosh, M.; Akbarizadeh, G. A subwavelength graphene surface plasmon polariton-based decoder. *Diam. Relat. Mater.* **2023**, *134*, 109780. [CrossRef]
20. Maleki, M.J.; Soroosh, M.; Akbarizadeh, G. A compact low-loss 2-to-4 plasmonic decoder based on suspended graphene for surface plasmon polariton transmission. *Diam. Relat. Mater.* **2024**, *144*, 110983. [CrossRef]
21. Han, L.; Chen, J.; Su, H.; Wang, H.; Wang, S.; Li, Q. Terahertz 2–4 decoder based on a dual-peak electromagnetically induced transparency metasurface. *Phys. Scr.* **2025**, *100*, 095511. [CrossRef]
22. Bao, G.; Li, P. *Maxwell's Equations in Periodic Structures*, 1st ed.; Springer: Singapore, 2022. [CrossRef]
23. Maleki, M.J.; Soroosh, M. A low-loss subwavelength plasmonic waveguide for surface plasmon polariton transmission in optical circuits. *Opt. Quantum Electron.* **2023**, *55*, 1266. [CrossRef]
24. Bahadori-Haghighi, S.; Ghayour, R.; Sheikhi, M.H. Design and analysis of low loss plasmonic waveguide and directional coupler based on pattern-free suspended graphene sheets. *Carbon* **2018**, *129*, 653–660. [CrossRef]
25. Mohammadi, G.; Orouji, A.A.; Danaie, M. Highly selective single-mode graphene bandpass filter based on Wilkinson power divider structure. *Diam. Relat. Mater.* **2024**, *145*, 111141. [CrossRef]
26. Liu, Y.; Zhong, R.; Ding, H.; Liu, S. Graphene loaded double ridge plasmon Terahertz waveguide. *Eur. Phys. J. D* **2017**, *71*, 83. [CrossRef]
27. Zhao, W.; Li, H.; Xiao, X.; Jiang, Y.; Watanabe, K.; Taniguchi, T.; Zettl, A.; Wang, F. Nanoimaging of low-loss plasmonic waveguide modes in a graphene nanoribbon. *Nano Lett.* **2021**, *21*, 3106–3111. [CrossRef]
28. Hadi, M.M.; Mohammed, J.K.; Atiyah, S.M. Design of 4 × 2 optical encoder utilizing nano-structure plasmonic IMI waveguides. *Opt. Contin.* **2024**, *3*, 368–378. [CrossRef]
29. Ma, J.; Zeng, D.; Yang, Y.; Pan, C.; Zhang, L.; Xu, H. A review of crosstalk research for plasmonic waveguides. *Opto-Electron. Adv.* **2019**, *2*, 180022. [CrossRef]
30. Nayyeri Raad, A.; Saghaei, H.; Mehrabani, Y.S. An optical 2-to-4 decoder based on photonic crystal X-shaped resonators covered by graphene shells. *Opt. Quantum Electron.* **2023**, *55*, 452. [CrossRef]
31. Li, Z.; Zhu, A.; Hou, W.; Yang, X.; Cheng, L.; Hu, C.; Qiao, F.; Xing, L.; Mahapatra, R.N. A 2 to 4 graphene metamaterial decoder and tetrahydrofuran detection. *IEEE Sens. J.* **2024**, *24*, 28753–28760. [CrossRef]
32. Turki, M.F.; Al-Tamimi, H.M. Design of 2 × 4 optical decoder based on IMI plasmonic waveguide with slot. *J. Opt.* **2024**, 1–9. [CrossRef]
33. Wang, G.; Feng, W. On-chip Mach-Zehnder interferometer sensor with a double-slot hybrid plasmonic waveguide for high-sensitivity hydrogen detection. *Opt. Express* **2023**, *31*, 39500–39513. [CrossRef]
34. Butt, M.A. Loop-terminated Mach–Zehnder interferometer integrated with functional polymer for CO_2 gas sensing. *Appl. Sci.* **2024**, *14*, 4714. [CrossRef]
35. Torcal-Milla, F.J.; Lobera, J.; Roche, E.M.; Lopez, A.M.; Palero, V.; Andres, N.; Arroyo, M.P. Modified Mach–Zehnder interferometer for spatial coherence measurement. *Opt. Lett.* **2023**, *48*, 3127–3130. [CrossRef]
36. Doughan, I.; Oyemakinwa, K.; Ovaskainen, O.; Roussey, M. Strip-loaded Mach–Zehnder interferometer for absolute refractive index sensing. *Sci. Rep.* **2024**, *14*, 3064. [CrossRef]
37. Gangwar, R.K.; Pathak, A.K.; Kumar, S. Recent progress in photonic crystal devices and their applications: A review. *Photonics* **2023**, *10*, 1199. [CrossRef]

38. Maleki, M.J.; Mir, A.; Soroosh, M. Ultra-fast all-optical full-adder based on nonlinear photonic crystal resonant cavities. *Photonic Netw. Commun.* **2021**, *41*, 93–101. [CrossRef]
39. Askarian, A.; Parandin, F. Demonstration of an all-optical 2-bit digital-to-analog converter based on photonic crystal waveguides. *Opt. Contin.* **2025**, *4*, 1800–1809. [CrossRef]
40. Liu, L.; Ma, C.; Ye, M.; Yu, Z.; Xue, W.; Hu, Z.; Li, J. Photonic crystal nanobeam cavity with a high experimental Q factor exceeding two million based on machine learning. *J. Light. Technol.* **2022**, *40*, 7150–7158. [CrossRef]
41. Nuzaihan, M.N.; Hashim, U.; Md Arshad, M.K.; Rahim Ruslinda, A.; Rahman, S.F.A.; Fathil, M.F.M.; Ismail, M.H. Top-down nanofabrication and characterization of 20 nm silicon nanowires for biosensing applications. *PLoS ONE* **2016**, *11*, e0152318. [CrossRef]
42. Cardenas, J.; Poitras, C.B.; Robinson, J.T.; Preston, K.; Chen, L.; Lipson, M. Low loss etchless silicon photonic waveguides. *Opt. Express* **2009**, *17*, 4752–4757. [CrossRef]
43. Bolotin, K.I.; Sikes, K.J.; Jiang, Z.; Klima, M.; Fudenberg, G.; Hone, J.; Kim, P.; Stormer, H.L. Ultrahigh electron mobility in suspended graphene. *Solid State Commun.* **2008**, *146*, 351–355. [CrossRef]

Disclaimer/Publisher's Note: The statements, opinions and data contained in all publications are solely those of the individual author(s) and contributor(s) and not of MDPI and/or the editor(s). MDPI and/or the editor(s) disclaim responsibility for any injury to people or property resulting from any ideas, methods, instructions or products referred to in the content.

crystals

MDPI

Article

Low-Frequency Noise Characteristics of Graphene/h-BN/Si Junctions

Justinas Glemža [1], Ingrida Pliaterytė [1], Jonas Matukas [1], Rimantas Gudaitis [2], Andrius Vasiliauskas [2], Šarūnas Jankauskas [2] and Šarūnas Meškinis [2,*]

[1] Institute of Applied Electrodynamics and Telecommunications, Vilnius University, Saulėtekio 3, LT-10257 Vilnius, Lithuania; justinas.glemza@ff.vu.lt (J.G.); ingrida.pliateryte@ff.stud.vu.lt (I.P.); jonas.matukas@ff.vu.lt (J.M.)

[2] Institute of Materials Science, Kaunas University of Technology, K. Baršausko 59, LT-51423 Kaunas, Lithuania; rimantas.gudaitis@ktu.lt (R.G.); andrius.vasiliauskas@ktu.edu (A.V.); sarunas.jankauskas@ktu.lt (Š.J.)

* Correspondence: sarunas.meskinis@ktu.lt

Abstract: Graphene/h-BN/Si heterostructures show considerable potential for future use in infrared detection and photovoltaic technologies due to their adjustable electrical behavior and well-matched interfacial structure. The near-lattice match between graphene and hexagonal boron nitride (h-BN) enables the deposition of low-defect-density graphene on h-BN surfaces. This study presents a thorough exploration of the low-frequency electrical noise behavior of graphene/h-BN/Si heterojunctions under both forward and reverse bias conditions at room temperature. Graphene nanolayers were directly grown on h-BN films using microwave plasma-enhanced CVD. The h-BN layers were formed by reactive high-power impulse magnetron sputtering (HIPIMS). Four h-BN thicknesses were examined: 1 nm, 3 nm, 5 nm, and 15 nm. A reference graphene/Si junction (without h-BN) prepared under identical synthesis conditions was also studied for comparison. Low-frequency noise analysis enabled the identification of dominant charge transport mechanisms in the different device structures. Our results demonstrate that grain boundaries act as dominant defects contributing to increased noise intensity under high forward bias. Statistical analysis of voltage noise spectral density across multiple samples, supported by Raman spectroscopy, reveals that hydrogen-related defects significantly contribute to $1/f$ noise in the linear region of the junction's current–voltage characteristics. This study provides the first in-depth insight into the impact of h-BN interlayers on low-frequency noise in graphene/Si heterojunctions.

Keywords: defects; fluctuations; directly synthesized graphene/h-BN/Si junction; hexagonal boron nitride; microwave PECVD; high power impulse magnetron sputtering; low-frequency noise; Raman spectroscopy

1. Introduction

Nearly two decades after its initial isolation from graphite, graphene remains among the most intensively investigated materials across contemporary physical sciences, nanotechnology, and materials research. It has attracted significant attention as a key material for future electronic and photonic technologies owing to a unique combination of high optical transparency and electrical conductivity, superior charge carrier mobility, and outstanding thermal conductivity [1–4].

Importantly, heterostructures combining graphene with conventional semiconductors of IV and III–V groups have been realized [3,5–10]. Benefiting from the mature and scalable

https://doi.org/10.3390/cryst15090747

silicon semiconductor technology, graphene/Si heterojunctions exhibit promising potential for solar cells, broadband photodetectors, and gas sensing applications [1,11–17].

A notable advantage of these junctions is their ability to enhance silicon solar cells' sub-bandgap infrared conversion efficiency [18] and extend the operating band of Si-based photonics into the mid-infrared spectral range [16,19,20]. Despite the potential, graphene/Si junctions still face unresolved issues such as variability in ideality factors, Schottky barrier height inhomogeneity, and limited performance due to various interface defects, such as dangling bonds at the silicon surface [11,12,14,17,21].

Due to its sensitivity to structural imperfections and defects introduced during sample fabrication or device operation, low-frequency noise spectroscopy serves as a non-invasive and powerful diagnostic approach for characterizing electronic and optoelectronic devices [22–24]. Different types of noise can be categorized, including thermal (also known as Johnson or Nyquist) noise, which is caused by the random motion of charge carriers and has a "white" spectrum, i.e., is frequency independent. Shot noise is also characterized by a "white" spectrum and results from uncorrelated and stochastic transitions of charge carriers through a potential energy barrier.

Generation-recombination noise, which is distinguished by a Lorentzian-type spectrum, and flicker or $1/f$ noise arise from fluctuations in the conductance [23,25,26]. Among these, $1/f$ noise is particularly important for assessing device reliability and gaining insight into carrier transport mechanisms [22,23,26].

In practical applications, low-frequency fluctuations can fundamentally limit sensor sensitivity. The $1/f$ noise in graphene is generally attributed to correlated variations in the number of charge carriers and their mobility [26–30]. Flicker noise in graphene-based semiconductor devices is under intensive investigation because of its impact on real-world device performance [3,6,27,28,30–34]. However, only limited studies have specifically investigated low-frequency noise characteristics in graphene/silicon junctions [3,6,35]. These works validate the significance of $1/f$ noise, which is sometimes omitted and only shot noise is considered, in estimating photodetector parameters such as noise equivalent power. Together with Raman spectroscopy, low-frequency noise investigation is a powerful tool not only for defect density evaluation [35] but also for defect identification as this study aims to demonstrate.

The defect density in graphene is highly sensitive to the synthesis method [36]. Graphene can be directly deposited on Si substrates without the use of catalysts, employing plasma-enhanced chemical vapor deposition (PECVD). The key benefit of this approach is avoiding lengthy transfer procedures as well as associated contamination and mechanical damage [37]. However, such directly grown graphene is usually nanocrystalline and contains a high density of grain-boundary defects [38,39]. These defects can significantly limit the performance and reliability of devices and sensors based on directly synthesized graphene [40]. While grain boundaries as noise sources have been investigated earlier in graphene transistors [30], this study also reveals their impact on the rapid noise increase in graphene/Si structures with an inserted hexagonal boron nitride interlayer. Careful substrate selection can substantially reduce the density of grain boundaries and other defects, thereby improving material quality.

The near-lattice match between graphene and hexagonal boron nitride (h-BN) enables epitaxial growth of low-defect-density graphene [41–44]. It is noteworthy that graphene/silicon solar cell conversion efficiency may be raised more than three times by inserting an h-BN interlayer [1,13]. However, to date, graphene epitaxy on h-BN has primarily been employed using exfoliated hexagonal boron nitride flakes or h-BN films synthesized by CVD on Cu foils. Several recent studies described graphene synthesis on h-BN films directly deposited on substrates commonly used for semiconductor device fabrication, such as silicon or SiO_2; however, lower defect density in graphene was not achieved [4,45]. Despite this, significant improvements in the graphene/h-BN/silicon devices' photoelectric and photovoltaic properties were found compared to the graphene/Si junction [4].

Notably, a graphene-channel field-effect transistor (FET) fabricated on a boron nitride substrate and encapsulated by another boron nitride layer exhibited ultralow low-frequency noise [32]. Noise in such a device was much lower compared to the graphene-based FET fabricated on the commonly used SiO_2 substrate [33]. Nevertheless, the effects of the h-BN interlayer on the low-frequency noise properties of the graphene/monocrystalline silicon heterojunction have not been studied to date. Only Ui Yeon Won et al. [46] applied noise measurements to evaluate the detectivity of the graphene/h-BN/silicon photosensor, fabricated using transferred graphene and boron nitride flakes. However, no detailed analysis of the noise behavior was reported. In such structures, determining conductivity only from temperature-dependent current–voltage characteristics can be challenging, as the dominant conduction mechanism may vary with temperature or involve multiple competing mechanisms. Here, low-frequency noise measurements can provide valuable complementary information. Based on low-frequency noise analysis, only Schottky emission was identified at the graphene/Si junctions [3].

To address the knowledge gap, this study systematically investigates low-frequency fluctuations in graphene/h-BN/Si junctions and compares them with those in conventional graphene/Si diodes. The effects of the h-BN interlayer thickness and the structural features of graphene grown on h-BN films are analyzed. Our results demonstrate that both grain boundary and on-site defects significantly affect low-frequency noise behavior. Low-frequency noise spectroscopy was used to determine dominant conduction mechanisms at room temperature, avoiding the complexity of analyzing temperature-dependent current–voltage characteristics (this is illustrated in Supplement S3 "Identification of Schottky and Poole–Frenkel emissions via current–voltage characteristics"). This study provides the first in-depth insight into how h-BN interlayers influence low-frequency noise in graphene and Si heterostructures.

2. Materials and Methods

The samples were prepared on single-crystalline, double-side-polished, n-type silicon (100) wafers (Sil'tronix Silicon Technologies, Archamps, France). Their resistivity values were from 1 to 10 Ω·cm. High-power impulse magnetron sputtering (HIPIMS) was applied to deposit h-BN films onto the wafers. The boron nitride films deposition conditions can be seen in Table 1. Details of the interlayer deposition parameters are summarized in Table 1. Boron nitride films growth conditions were selected considering our previous study [47]. Detailed information on the structure and composition of the h-BN films utilized in this research is available in [47]. Notably, our previous study found that the B:N ratio for thicker h-BN films was 1:1 for 30 min to grow a film of ~50 nm thickness and 1.17 for 60 min to grow a film of ~100 nm thickness [47]. The continuous h-BN film was ultra-smooth with a roughness (Rq) of ~0.5 nm [4].

Graphene was synthesized via microwave PECVD using a Cyrannus system (Innovative Plasma Systems (Iplas) GmbH, Troisdorf, Germany). To protect the samples from excessive exposure to direct plasma, a specialized enclosure was utilized [35,48,49]. Before the graphene deposition, the substrate was cleaned by hydrogen plasma activation. The substrate pre-treatment and graphene synthesis conditions are in Table 2. It is worth mentioning that our previous research indicated that the roughness of graphene grown on a continuous h-BN interlayer was below 1 nm [4].

Table 1. h-BN film deposition settings.

Target	Deposition Temperature (°C)	Impulse On-time (t_{on}) (μs)	Impulse Off-time (t_{off}) (μs)	Average Impulse Current (A)	Average Voltage (V)	Substrate-Cathode Distance (cm)	Interlayer Thicknesses (nm)
Boron	800	17	150	1.2	930	15	1, 3, 5, 15

Table 2. Graphene synthesis process technological parameters.

Technological Process	Plasma Power (kW)	Temperature (°C)	H_2 Gas Flow (sccm)	CH_4 Gas Flow (sccm)	Pressure (mBar)	Time (min)
Graphene pre-treatment	1	700	200	0	10	10
Graphene growth	0.7	700	75	25	10	60

Diodes based on graphene, hexagonal boron nitride, and silicon were fabricated to assess their electrical and low-frequency noise properties. The cross-sectional view of the device is shown in Figure 1. For more details on diode fabrication, refer to previous work [4].

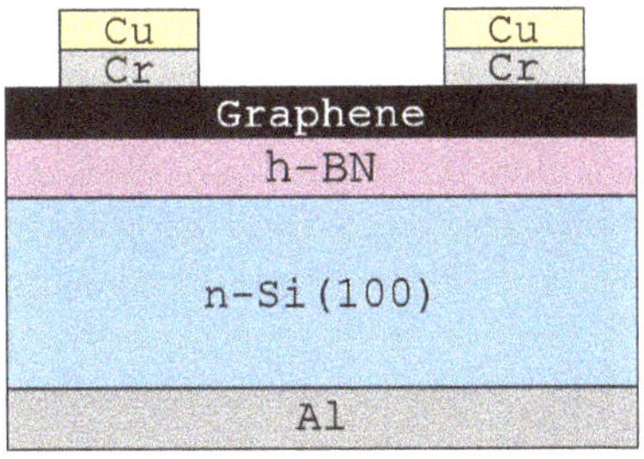

Figure 1. Schematic of the graphene/h-BN/Si(100) junction device.

The thickness of boron nitride films was estimated based on the deposition rate of the thicker films revealed in our previous research [47] and the corresponding growth duration.

Low-frequency noise measurements were conducted across the 10 Hz to 60 kHz frequency band under constant current conditions, for both forward and reverse bias, at ambient temperature. The noise measurement setup consists of a specially designed low-noise amplifier, filtering modules, and analog-to-digital converter PCI-6115 (National Instruments, Austin, TX, USA). The spectral density of voltage fluctuations' (S_U) is derived by applying the Cooley–Tukey Fast Fourier Transform algorithm. All noise measurements were carried out inside a Faraday cage to minimize electromagnetic interference. A detailed noise measurement description and a schematic of the experimental setup (Figure S1) are provided in Supplement S1. Four reference graphene/Si junctions and six graphene/h-BN/Si samples were tested within each h-BN thickness category. A B1500A semiconductor parameter analyzer (Keysight, Santa Rosa, CA, USA) with minimum current measurement resolution of 100 fA was employed to acquire current–voltage (*I-V*) characteristics. *I-V* characteristics under CW operation were measured with a voltage step of 10 mV with 0 ms delay time at forward and reverse sweeps starting at 0 V. The compliance DC current was limited to 20 mA and the maximum voltage to ±5 V to avoid the device under investigation damage. The room temperature maintained in the laboratory was 295 K.

The structural properties of the samples were analyzed with the aid of an inVia (Renishaw, United Kingdom) Raman spectrometer. The exciting laser wavelength of 532 nm was applied. We used a laser power of 1.5 mW for measurements. Raman spectra were recorded from five locations on each sample. Deconvolution of the 2D, G, D, and D′ bands was conducted using XPSPeak 4.1 software, fitting them with Lorentzian line shapes. The 2D peak and G peak intensity ratio (I(2D)/I(G)) was employed to appraise the number of layers [50]. A reduction in this ratio implies an increase in the number of graphene layers. The ratio of the D peak intensity to the G peak intensity (I(D)/I(G)) was utilized to estimate the defect density within the graphene structure [51,52]. An increased ratio reflects a higher density of defects.

3. Results

3.1. I-V Characteristics

Typical current–voltage characteristics of investigated junctions with different h-BN layer thicknesses at room temperature are shown in Figure 2 (these current–voltage characteristics in semi-log scale are presented in Figure S2 in Supplement S2). No apparent differences are observed between the samples with a h-BN layer and without. At forward bias in the low-voltage regime and at reverse bias over a broader range, the current scales linearly with voltage, indicating dominant ohmic behavior. At higher forward voltages, a nonlinearity appears in all samples, indicating possible Schottky emission or other field-enhanced conduction mechanisms, such as Poole–Frenkel (PF) emission, direct tunneling, or Fowler–Nordheim tunneling. Such conduction mechanisms are often observed in graphene-based devices [11,17,53]. In graphene/semiconductor junctions, Schottky emission is widely accepted as a dominant conduction mechanism [3,11,16,17]. However, when an insulator, such as h-BN, is inserted between graphene and a semiconductor, the determination of current conduction requires special attention. For example, when direct tunneling is dominant, the Richardson coefficient in the Schottky equation should be modified with a transmission factor. This factor depends on the insulator thickness and the barrier height at the semiconductor/insulator interface [1,11].

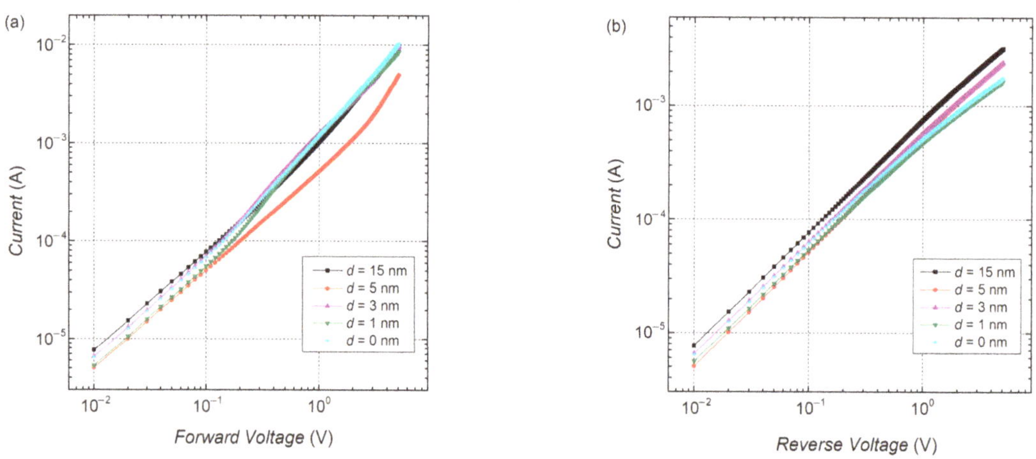

Figure 2. Typical current–voltage characteristics of the studied junctions at forward (**a**) and reverse (**b**) bias; d—h-BN interlayer thickness.

Additionally, in such graphene/insulator/semiconductor structures, multiple conduction mechanisms can be active simultaneously or the dominant conduction mechanism can change with temperature. It should be noted that inhomogeneities at the interface, defects, conduction at grain boundaries, and barrier height fluctuations complicate the accurate identification of dominant conduction mechanisms [12,17,54]. The complexity in determining the Schottky vs. Poole–Frenkel emission from *I-V* characteristics measurement in our investigated samples is described in Supplement S3. That is where noise analysis helps. Ln(I) vs. $\sqrt{U}$ and ln(I/U) vs. $\sqrt{U}$ plots, which are usually employed for Schottky and Poole–Frenkel emission identification, are also presented in Figures S3 and S4. In the following section, the dominant conduction mechanism will be estimated based on the noise characteristics.

3.2. Low-Frequency Noise Spectra

All investigated samples exhibit $1/f$ or $1/f^{\alpha}$-type electrical fluctuations at forward and reverse bias. These fluctuations can be described as a superposition of many Lorentzian-type noise spectral components resulting from individual fluctuators with widely distributed relaxation times. The representative noise spectra of the samples at forward and reverse bias are shown in Figure 3. Some junctions containing h-BN interlayers, investigated in this study, are distinguished by the Lorentzian-type noise spectra superimposed with $1/f$ components at low frequencies at forward bias at a specific range of flowing current. Several examples of electrical noise spectra with Lorentzian fluctuations are presented in Figure 4. In samples with a 5 nm thickness h-BN interlayer, Lorentzian-type noise spectra have been observed most frequently: in the current span from 0.05 mA to 0.1 mA and at higher currents from 0.75 mA to 3 mA. Additionally, a weak Lorentzian-type noise component has been observed in a junction with a 1 nm h-BN layer at 0.2 mA forward current (I_F) (Figure 4). To facilitate clearer recognition of Lorentzian-type fluctuations, these spectra are commonly scaled by multiplying with frequency (S_{Uf}) (Supplement S4, Figure S5). The Lorentzian-type noise spectrum is indicative of an isolated defect-related recombination site with a distinct relaxation time $1/\tau_r = 1/\tau_e + 1/\tau_c$, where τ_e is the average time of electron emission, and τ_c is the average electron capture time at the trap level [24]. In general, the current flow through the junction results in voltage noise due to variations in local resistance [25,55,56]. If the number of defects (both bulk and interface types) in the device is sufficiently high and the relaxation times are widely distributed, the $1/f$ noise spectrum is observed. The equation for $1/f$ noise intensity evaluation (Equation S4) and its description are provided in Supplement S4.

A "white" spectral region attributed to thermal noise is observed at low forward and reverse bias when $f > 2$ kHz, as can be seen in Figure 3 at 0.005 mA and 0.01 mA currents. At such low currents, thermal noise is dominant and sets a limit for the minimum observable noise level. The intensity of thermal noise ($S_U = 4kTR_d$, k—Boltzmann's constant, T—temperature) is determined by the differential resistance R_d of the investigated junction at the linear *I-V* characteristic part (for a representative sample of a 3 nm thickness h-BN layer in Figure 3, $R_d = 1410\ \Omega$). In some cases, thermal noise is masked by higher $1/f$ noise intensity. An example is illustrated in Figure 3b for a junction with a 5 nm h-BN layer at 0.01 mA.

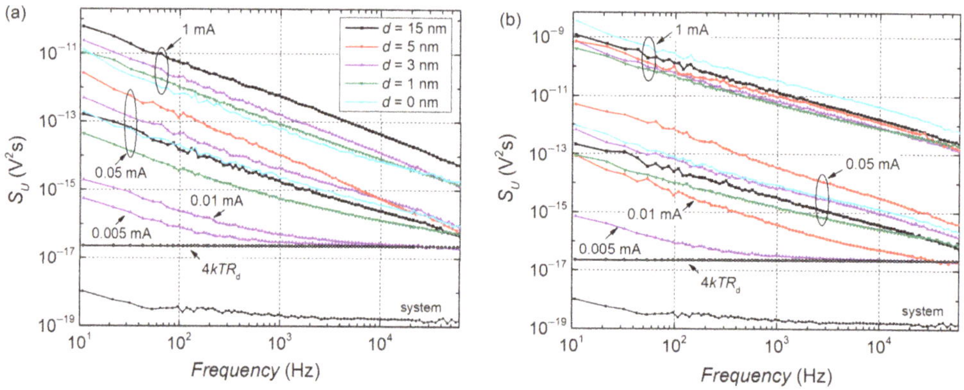

Figure 3. Representative spectra of low-frequency noise at different values of the forward (**a**) and reverse (**b**) current. The curve named "system" denotes the noise level of the measurement system; d—h-BN interlayer thickness. The legend is the same for (**a**,**b**) figures.

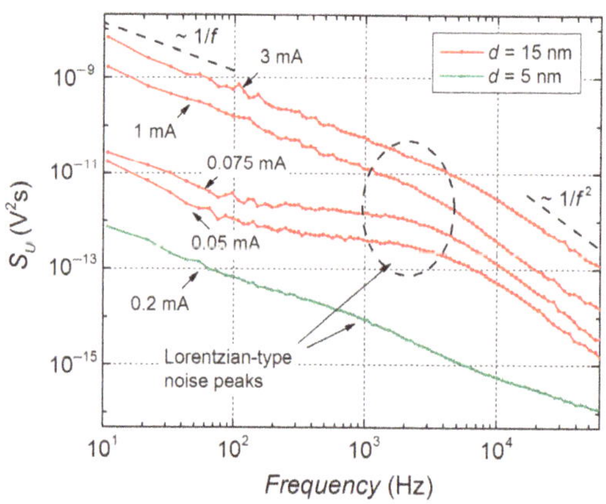

Figure 4. Electrical noise spectra with Lorentzian-type components at different forward current values; d—h-BN interlayer thickness.

3.3. Noise Spectral Density vs. Current Characteristics

The interdependence of the current noise spectral density S_I and the current I provides insight into the physical processes occurring at the junction and possible conduction mechanisms [3,23,35,57]. For this reason, the current noise spectral density has been calculated as $S_I = S_U/R_d^2$, where R_d denotes the differential resistance of the sample $R_\mathrm{d} = (\mathrm{d}U/\mathrm{d}I)$. The current dependences of the differential resistance for representative samples shown in Figure 2 are depicted in Figure S6 (Supplement S5).

3.3.1. Influence of Grain Boundaries on Low-Frequency Noise

Figure 5 presents the different trends observed in the current spectral density for forward and reverse current in the samples investigated. In the current range, in which the I-V characteristic is linear, electrical noise S_I is proportional to I^γ, where γ is very close to 2 for all the investigated junctions. This is expected for ohmic samples, as Equation (S4) defines. At elevated forward bias, a discrepancy from the quadratic relationship in noise vs. current characteristic is observed, where $S_I \sim I^\gamma$, and $\gamma > 2$. The current intervals with a rapid increase in noise intensity coincide with the non-linear regions of the I-V characteristics. Such strong noise dependence on current with $\gamma > 2$ is often attributed to the influence of grain boundaries (GB) and was earlier observed in polycrystalline films [23]. As a result of the nanocrystalline structure of graphene [39,48,58,59] and h-BN layers [45], grain boundaries are dominant defect sources in these devices. Accordingly, in the investigated graphene/h-BN/Si junctions, the observed $S_I \sim I^3$ behavior confirms the dominant role of GBs as strong noise generators [30]. In graphene grown by CVD, GB prevailing results in a substantial increase in noise intensity compared to exfoliated graphene [36].

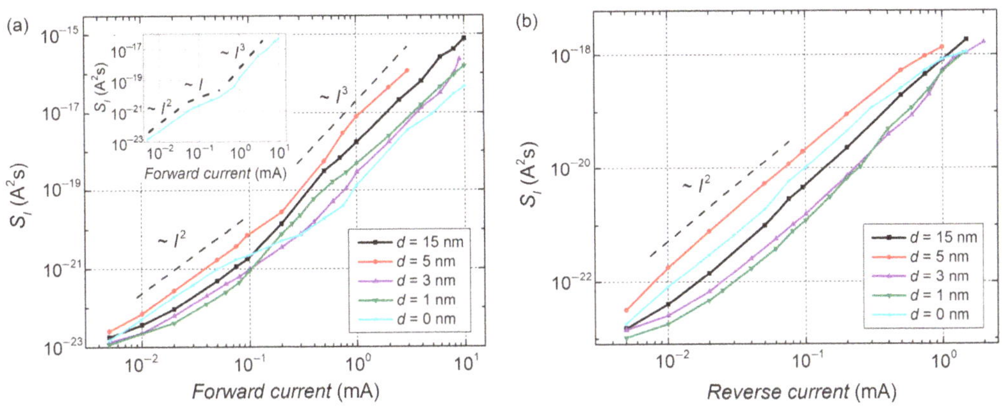

Figure 5. Typical graphs of current noise spectral density vs. forward (**a**) and reverse (**b**) current at 1044 Hz frequency; *d*—h-BN interlayer thickness. The inset in (**a**) represents the $S_I(I)$ dependence of the reference sample.

However, reference samples without an h-BN layer are distinguished by different electrical noise vs. forward current characteristics (Figure 5a). An interval, in which S_I scales approximately linearly with current (i.e., $\gamma \approx 1$) is detected within the current range of 0.05 to 0.4 mA. At high currents, the influence of GBs plays a significant role in the rapid increase in noise intensity. The proportionality $S_I \sim I$ am usually attributed to pure thermionic emission in Schottky diodes [3,57]. Such a dependence was not observed in samples with an h-BN layer, meaning that a different prevailing conduction mechanism or several conduction mechanisms are contributing simultaneously.

3.3.2. Identification of Poole–Frenkel Emission

Following the approach outlined in [60–62] for dominant Poole–Frenkel emission, the product of normalized current noise spectral density S_I/I^2 and electrical field (which is proportional to the applied voltage U) should be constant regardless of the current and is proportional to the trap density and field enhancement factor of PF emission. A plateau in corresponding noise plots is observed at high reverse bias for samples with 15 nm and 5 nm thicknesses of the h-BN layer, suggesting the dominance of Poole–Frenkel emission. A trend has been observed—samples with thicker h-BN layers exhibit higher Poole–Frenkel conductivity at room temperature, as shown in Figure 6. The PF noise analysis at high forward bias was complicated by the influence of grain boundaries on noise characteristics. Only one sample with a 15 nm thickness h-BN layer demonstrated a narrow plateau in $U \times S_I/I^2$ vs. current plot at high forward bias (Figure 6a), while no such region was observed in the 5 nm thickness h-BN samples. The PF charge transport mechanism has been reported for a small forward voltage range of the *I-V* characteristics of graphene/Al_2O_3/Si solar cells [11], a high reverse voltage range of the *I-V* characteristics of graphene/n-Si junctions [21], and reverse current flow via graphene/SiC junctions [9].

The Poole–Frenkel emission from defects in h-BN was found in graphene/h-BN field effect transistors [10]. Similarly, in the present study, it is plausible that PF conduction arises from defects in the HIPIMS-deposited h-BN interlayer. Exfoliated h-BN layers or h-BN films grown by chemical vapor deposition on catalytic copper foil usually contain few defects, and their use can ensure the fabrication of graphene-based FETs with ultra-low noise. However, these technologies are not suited for industrial-scale semiconductor device fabrication. In our case, the boron nitride was grown directly on a silicon substrate using the HIPIMS technique, which is already used for large-scale manufacturing. In this case, higher defect density in the h-BN films can be supposed like [63].

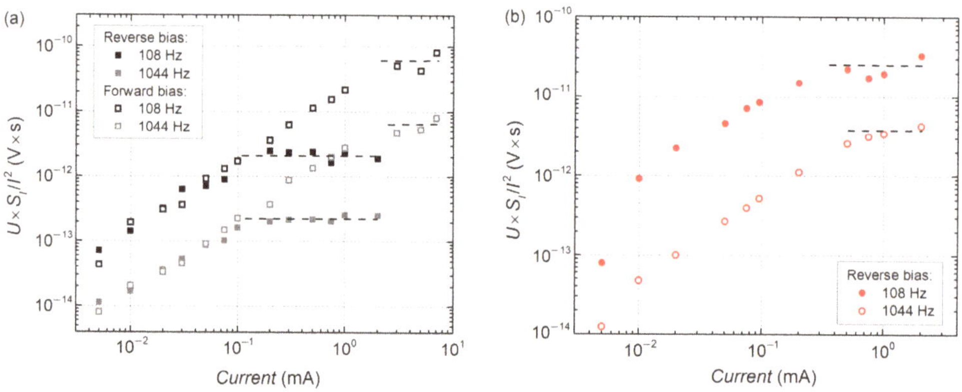

Figure 6. $U \times S_I/I^2$ vs. current plots for 15 nm (**a**) and 5 nm (**b**) h-BN layer thickness samples at 108 Hz and 1044 Hz frequencies.

From the low-frequency noise perspective, the pronounced dominance of trap-assisted tunneling is usually linked to the appearance of shot noise in electrical noise spectra, modulated by the Fano factor [61]. However, no shot noise was observed in this experiment.

Here, in Figures 3–5, only representative noise curves or dependences of samples are represented. Direct comparison of noise intensity between different samples should rely on Figure 7a in the following Section 3.4, which summarizes the averaged data.

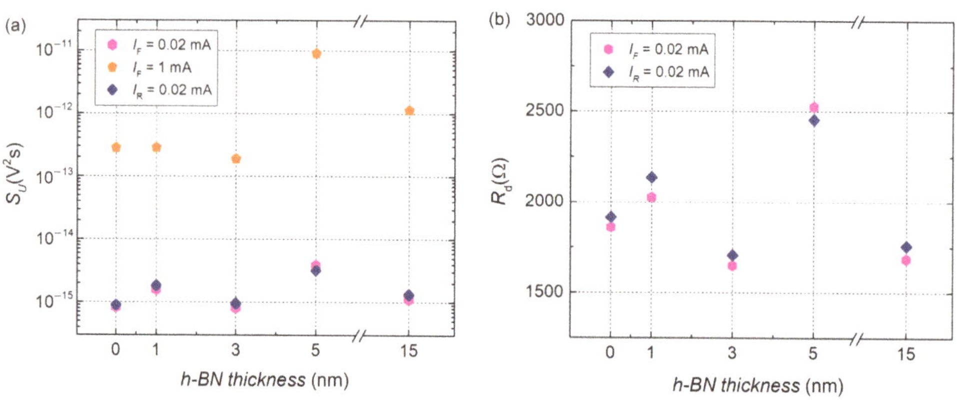

Figure 7. Average voltage noise spectral density values at a 1044 Hz frequency, reverse current of 0.02 mA, and forward current of 0.02 and 1 mA (**a**); differential resistance in the ohmic part of the current–voltage curve under positive and negative biasing conditions (**b**) of multiple graphene/h-BN/Si junction samples as a function of h-BN layer thickness.

3.4. Statistical Investigation and Defect Identification in Samples with Different H-BN Layer Thickness

3.4.1. Noise Intensity Between Samples with Different H-BN Layer Thickness

To compare the noise intensity between different samples and evaluate the impact of h-BN interlayer thickness on the graphene/h-BN/Si junctions through the low-frequency noise characterization, voltage noise spectral density values at 1044 Hz were extracted for multiple investigated samples within each h-BN thickness category. The average values of voltage noise spectral densities of the tested junctions are presented in Figure 7a. The comparison was made at different forward current levels (0.02 mA and 1 mA), corresponding to the linear and non-linear regions of the *I-V* characteristics (Figure 2), as well as at 0.02 mA of reverse bias. The noise intensity is not directly proportional to the h-BN thickness (Figure 7a). As a general trend, junctions with a 5 nm-thick h-BN interlayer demonstrated the highest noise intensity under all experimental conditions. The average voltage noise spectral density is very similar between specimens with h-BN interlayers under both forward and reverse biases in the linear *I-V* characteristic range, except for graphene/5 nm h-BN/Si samples (Figure 7a). However, under forward bias conditions, where the *I-V* behavior deviates from linearity, the fluctuation intensity of the thickest h-BN layer samples (5 nm and 15 nm) increases more compared to the increase in noise level of the 1 nm and 3 nm h-BN thickness samples (I_F = 1 mA in Figure 7a).

This observation is consistent with the findings in Section 3.3, where a deviation from the $S_I \sim I^2$ dependence and an enhanced $\gamma > 2$ exponent were linked to grain boundaries (GBs) acting as dominant noise sources in nanocrystalline graphene and h-BN layers. These GB-related effects can also contribute to local current crowding and a further increase in $1/f$ noise [23,31]. Thus, it is important to consider in which mode (linear or non-linear *I-V* characteristic part) the device operates. As Figure 7b illustrates, a good correlation between the noise intensity and the differential resistance values of the investigated junctions in the linear *I-V* characteristic range is confirmed.

The reference junctions without h-BN interlayers are characterized by close (or even slightly lower) noise intensities as the remaining samples (except the graphene/5 nm h-BN/Si junctions) (Figure 7a). The values of differential resistances are also similar to the remaining samples (Figure 7b).

3.4.2. Assessment of Raman Spectra and Low-Frequency Noise Characteristics

Raman scattering spectra were studied to reveal structural parameters of the graphene and relate them to the low-frequency noise results. G and 2D peaks characteristic of graphene are apparent in those spectra (Supplement 6, Figure S7). The defect-related D band was found because directly deposited graphene is nanocrystalline, and another defect-related peak, D′ peak, is observed as a shoulder of the G band. Afterward, the Raman spectra parameters were analyzed to reveal possible relations between the low-frequency noise and graphene's structural parameters.

It was revealed in our previous study that substrate-induced graphene self-doping is crucial for graphene/Si junction properties and that the insertion of an h-BN interlayer can partially suppress this effect, thereby enhancing the photoelectric and photovoltaic properties of the junction [4]. Analysis of the Pos(2D) vs. Pos(G) plot can provide valuable information on stress and doping of the graphene [64]. The Pos(2D) downshift with Pos(G) upshift, typical of n-type doping, was revealed for graphene grown directly on silicon and on the h-BN interlayer [4]. However, in the present research, no dependence of low-frequency noise on Pos(G) or Pos(2D) has been observed (Supplement S6, Figure S8). This indicates that substrate-induced graphene self-doping should not be considered the primary factor determining the noise characteristics at graphene/h-BN/Si junctions. Notably, no apparent relationship between the spectral density of electrical noise and graphene's layer number (Supplement S6, Figure S9) has also been observed.

Non-monotonic noise increase in defect density is observed in graphene (Figure 8a). Initially, the noise clearly increases with the $I(\mathrm{D})/I(\mathrm{G})$ ratio. However, for samples featuring an inserted 1 nm-thick h-BN interlayer, a significant reduction in the noise level is observed. To identify possible reasons for this inconsistency, the effects of different graphene defect types were investigated. For this reason, intensity ratios of the D and D′ peaks were calculated. Three groups of the samples can be distinguished. In the case of the graphene/h-BN/Si junctions containing 3 nm, 5 nm, and 15 nm thickness boron nitride interlayers, the $I(\mathrm{D})/I(\mathrm{D'})$ value was about 2.5. That is between the value typical for grain boundary defects (3.5) and the value typical for on-site defects (1.3), which are commonly linked to hydrogen species bonded to the graphene [65]. In the case of the reference graphene sample, GB defects also prevail. However, considering these $I(\mathrm{D})/I(\mathrm{D'})$ values, the presence of some on-site defects is observed. Finally, for graphene/h-BN/Si(100) samples with an inserted 1 nm thickness h-BN interlayer, the $I(\mathrm{D})/I(\mathrm{D'})$ value appears to nearly match the typical value for grain boundaries (~3.6). It should be noted that the presence of some hydrogen bonds was revealed in HIPIMS-deposited h-BN films grown using a 152 sccm nitrogen gas flow using Raman spectroscopy [4]. In HIPIMS, hydrogen can originate from residual water vapor or hydrocarbons in the chamber; these species are dissociated in plasma and become bonded to boron or nitrogen at the growing film's surface, as was demonstrated in our previous research [4]. In the present study, such hydrogen incorporated in the h-BN layer could subsequently be released during the graphene's direct synthesis step and become incorporated into the growing graphene. For the 1 nm h-BN film, the absence of detectable hydrogen related features in the graphene suggests that either hydrogen was initially negligible, or it was fully desorbed during synthesis at 700 °C. For thicker h-BN interlayers, hydrogen atoms could diffuse from the boron nitride into the graphene layer during growth, becoming incorporated into its structure.

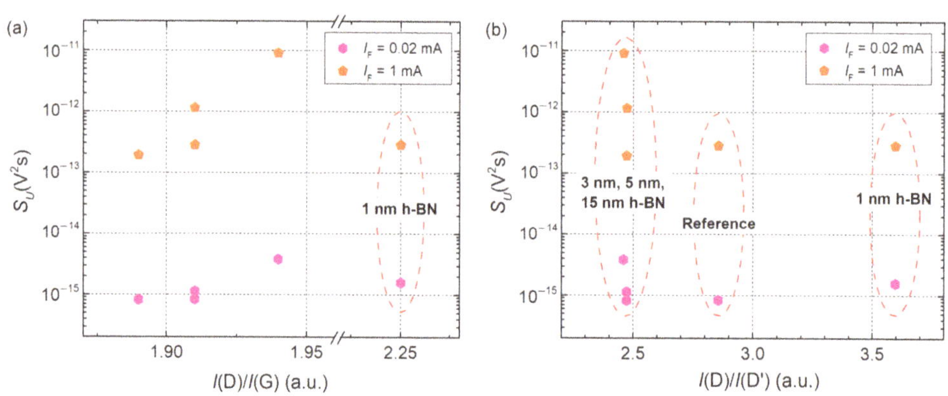

Figure 8. Average voltage noise spectral density vs. $I(\mathrm{D})/I(\mathrm{G})$ (**a**) and $I(\mathrm{D})/I(\mathrm{D'})$ ratios (**b**). The S_U was measured using an I_F of 0.02 mA and 1 mA and a frequency of 1044 Hz.

It should be noted that in the case of graphene/Si junction X-ray detectors and graphene-based field-effect transistors, hydrogen-related defects can lead to a significant increase in low-frequency noise [66,67]. Thus, the existence of hydrogen-related structural irregularities in the graphene may be an important factor affecting the noise increase in the device. This explains the outstanding samples with 5 nm thickness h-BN layer in Figure 7. On the other hand, in our case, average voltage noise spectral density in the graphene/1 nm h-BN/silicon junctions is not exceptionally large despite much higher grain boundary density indicated by high $I(D)/I(G)$ proportion (Figure 8a). This effect was previously reported during earlier $1/f$ noise studies on graphene/silicon contacts [35]. It was attributed to an increase in charge carriers at the graphene grain boundaries, which act as dopants with very low excitation energies [35].

In general, several tendencies have been observed when comparing the results obtained in Section 3.3 with the analysis of Raman spectra. For reference, 3 nm, 5 nm, and 15 nm thickness h-BN samples, it can be stated that hydrogen-related defects are mainly responsible for the fluctuations observed in the linear *I-V* characteristic part and have an influence on the differential resistance of the investigated junctions, as demonstrated by a 5 nm thickness h-BN sample. At higher forward currents, the uneven noise increases between junctions with different h-BN thicknesses, as well as the change in S_I dependence on current, indicates that the influence of other defects, primarily grain boundaries, increases in addition to hydrogen-related defects. Furthermore, the inserted h-BN layer affects the low-frequency fluctuations in the non-linear *I-V* region, especially in the case of a thick h-BN layer (Figure 7a), when the GBs start to dominate the conduction.

4. Conclusions

A comprehensive assessment of the low-frequency electrical noise characteristics of graphene/h-BN/Si heterojunctions has been performed under both forward and reverse bias and compared with the graphene/silicon reference structure without the h-BN interlayer.

All investigated junctions exhibit $1/f$-type electrical fluctuations at both forward and reverse bias, with the appearance of some Lorentzian-type noise components at specific forward current intervals. Low-frequency noise analysis indicates Schottky emission in reference graphene/Si junctions at room temperature. In contrast, samples with thicker h-BN interlayers (15 nm and 5 nm) exhibit signatures of Poole–Frenkel emission, with its contribution diminishing as the h-BN layer thickness decreases.

The statistical investigation of multiple samples across each h-BN layer thickness category, combined with analysis of graphene Raman spectra, reveals the presence of hydrogen-related defects in graphene and at the graphene/h-BN interface. These defects are responsible for the observed low-frequency fluctuations at low bias in the linear region of the device's current–voltage characteristic. Among the investigated junctions, the graphene/1 nm h-BN/Si structure showed a significantly reduced density of these defects.

A rapid increase in noise intensity in the non-linear junction *I-V* characteristic range, where current spectral density S_I scales as I^γ and $\gamma > 2$, is associated with the grain boundaries being dominant defects and acting as major noise sources due to the nanocrystalline structure of the graphene and h-BN layers.

Overall, the results demonstrate that h-BN interlayer thickness and defect composition play a critical role in governing low-frequency noise behavior, which is essential for optimizing graphene-based devices for sensing and optoelectronic applications.

Supplementary Materials: The following supporting information can be downloaded at: https://www.mdpi.com/article/10.3390/cryst15090747/s1, Figure S1. The noise measurement setup: A—an ammeter, ADC—an analog-to-digital converter, B—batteries, F—a filter system, LNA—a low-noise amplifier, PC—a personal computer, R_{load}—a load resistance, R_{ref}—a reference resistor, Sample—a graphene/h-BN/Si or graphene/Si junction under test, V—a voltmeter; Figure S2. Typical current–voltage characteristics of the investigated junctions at forward and reverse bias in semi-log scale; d—h-BN interlayer thickness; Figure S3. Plots of $\ln(I)$ vs. $\sqrt{U}$ (usually used for Schottky emission identification) for junctions with h-BN layer thickness of 15 nm (a) and 5 nm (b), and for the reference sample without the h-BN layer (c) at forward bias at different temperatures. Red lines are the linear fits at room temperature; Figure S4. Plots of $\ln(I/U)$ vs. $\sqrt{U}$ (usually used for Poole–Frenkel emission identification) for junctions with h-BN layer thickness of 15 nm (a) and 5 nm (b), and for the reference sample without the h-BN layer (c) at forward bias at different temperatures. Red lines are the linear fits at room temperature; Figure S5. Normalized electrical noise spectra (S_{Uf}) with Lorentzian-type components at different forward current values; *d*—h-BN interlayer thickness. The dashed area indicates where $1/f$ fluctuations dominate. The frequency of the maximum of the normalized voltage fluctuation spectral density f_0 denotes the relaxation time as $\tau_r = 1/(2\pi f_0)$; Figure S6. Dependences of the differential resistance on forward (a) and reverse (b) currents for samples shown in Figure 2; *d*—h-BN interlayer thickness; Figure S7. Typical Raman scattering spectra of the graphene samples; Figure S8. Dependences of average voltage noise spectral density on Pos(2D) (a) and Pos(G) (b) at forward currents of 0.02 mA and 1 mA at 1044 Hz frequency; Figure S9. Dependence of average voltage noise spectral density on $I(2D)/I(G)$ ratio at forward currents of 0.02 mA and 1 mA at 1044 Hz frequency; Table S1. Experimental slope values from linear fit in the high-bias regime at 295 K for graphene/h-BN/Si samples and the reference graphene/Si junction. References [9,25,55,56,68,69] are cited in the Supplementary Materials.

Author Contributions: Conceptualization, J.G., J.M. and Š.M.; methodology, J.M. and Š.M.; validation, J.G. and Š.J.; formal analysis, J.G. and I.P.; investigation, J.G., I.P., A.V., Š.J. and R.G.; resources, J.G., R.G. and A.V.; writing—original draft preparation, J.G., J.M. and Š.M.; writing—review and editing, J.G., J.M. and Š.M.; visualization, J.G., I.P. and Š.J.; supervision, J.M. and Š.M.; project administration, Š.M.; funding acquisition, Š.M. All authors have read and agreed to the published version of the manuscript.

Funding: This study was funded by the Research Council of Lithuania (Proposal No. P-MIP-22-235; Contract No. S-MIP-22-67).

Data Availability Statement: Data is contained within the article or supplementary material.

Conflicts of Interest: The authors declare no conflicts of interest.

Abbreviations

The following abbreviations are used in this manuscript:

h-BN	Hexagonal boron nitride
HIPIMS	High power impulse magnetron sputtering
PECVD	Plasma-enhanced chemical vapor deposition
CVD	Chemical vapor deposition
FET	Field-effect transistor
S_U	Spectral density of voltage fluctuations'
S_I	Spectral density of current fluctuations'
S_R	Spectral density of resistance fluctuations'
I-V characteristics	Current–voltage characteristics
R_d	Differential resistance
Pos(G)	Position of G peak
Pos(2D)	Position of 2D peak
PF emission	Poole–Frenkel emission
GB	Grain boundaries

References

1. Wang, C.; Behura, S.K.; Berry, V. Temperature Dependent Device Characteristics of Graphene/H-BN/Si Heterojunction. *Semicond. Sci. Technol.* **2020**, *35*, 075020. [CrossRef]
2. Shen, J.; Liu, X.; Song, X.; Li, X.; Wang, J.; Zhou, Q.; Luo, S.; Feng, W.; Wei, X.; Lu, S.; et al. High-Performance Schottky Heterojunction Photodetector with Directly Grown Graphene Nanowalls as Electrodes. *Nanoscale* **2017**, *9*, 6020–6025. [CrossRef]
3. An, Y.; Behnam, A.; Pop, E.; Bosman, G.; Ural, A. Forward-Bias Diode Parameters, Electronic Noise, and Photoresponse of Graphene/Silicon Schottky Junctions with an Interfacial Native Oxide Layer. *J. Appl. Phys.* **2015**, *118*, 114307. [CrossRef]

4. Meškinis, Š.; Jankauskas, Š.; Vasiliauskas, A.; Stankus, V.; Guobienė, A.; Lukoševičius, K.; Jasutis, A.; Gudaitis, R. Graphene Direct Growth by Microwave PECVD on H-BN Films Deposited by Reactive HIPIMS. *Nano Ex.* **2025**, *6*, 025007. [CrossRef]
5. An, Y.; Behnam, A.; Pop, E.; Ural, A. Metal-Semiconductor-Metal Photodetectors Based on Graphene/P-Type Silicon Schottky Junctions. *Appl. Phys. Lett.* **2013**, *102*, 013110. [CrossRef]
6. Zhu, M.; Li, X.; Li, X.; Zang, X.; Zhen, Z.; Xie, D.; Fang, Y.; Zhu, H. Schottky Diode Characteristics and $1/f$ Noise of High Sensitivity Reduced Graphene Oxide/Si Heterojunction Photodetector. *J. Appl. Phys.* **2016**, *119*, 124303. [CrossRef]
7. Kalita, G.; Shaarin, M.D.; Paudel, B.; Mahyavanshi, R.; Tanemura, M. Temperature Dependent Diode and Photovoltaic Characteristics of Graphene-GaN Heterojunction. *Appl. Phys. Lett.* **2017**, *111*, 013504. [CrossRef]
8. Li, H.; Shi, Y.; Shang, H.; Wang, W.; Lu, J.; Zakharov, A.A.; Hultman, L.; Uhrberg, R.I.G.; Syväjärvi, M.; Yakimova, R.; et al. Atomic-Scale Tuning of Graphene/Cubic SiC Schottky Junction for Stable Low-Bias Photoelectrochemical Solar-To-Fuel Conversion. *ACS Nano* **2020**, *14*, 4905–4915. [CrossRef]
9. Tomer, D.; Rajput, S.; Hudy, L.J.; Li, C.H.; Li, L. Carrier Transport in Reverse-Biased Graphene/Semiconductor Schottky Junctions. *Appl. Phys. Lett.* **2015**, *106*, 173510. [CrossRef]
10. Quezada-Lopez, E.A.; Joucken, F.; Chen, H.; Lara, A.; Davenport, J.L.; Hellier, K.; Taniguchi, T.; Watanabe, K.; Carter, S.; Ramirez, A.P.; et al. Persistent and Reversible Electrostatic Control of Doping in Graphene/Hexagonal Boron Nitride Heterostructures. *J. Appl. Phys.* **2020**, *127*, 044303. [CrossRef]
11. Wong, H.; Zhang, J.; Liu, J.; Anwar, M.A. On the Current Conduction and Interface Passivation of Graphene–Insulator–Silicon Solar Cells. *Nanomaterials* **2025**, *15*, 416. [CrossRef]
12. Tomer, D.; Rajput, S.; Hudy, L.J.; Li, C.H.; Li, L. Inhomogeneity in Barrier Height at Graphene/Si (GaAs) Schottky Junctions. *Nanotechnology* **2015**, *26*, 215702. [CrossRef]
13. Afzal, A.M.; Imran, M.; Iqbal, M.W.; Iqbal, M.Z.; Mumtaz, S.; Azeem, M.; Dastgeer, G.; Al-Ammar, E.A.; Ali, A. Surface Modification and Interface Engineering to Enhance the Performance of 2D-Graphene/3D-Silicon Schottky Junction Solar Cells. *J. Mater. Sci. Mater. Electron.* **2024**, *35*, 2165. [CrossRef]
14. Sinha, D.; Lee, J.U. Ideal Graphene/Silicon Schottky Junction Diodes. *Nano Lett.* **2014**, *14*, 4660–4664. [CrossRef] [PubMed]
15. Li, X.; Zhu, M.; Du, M.; Lv, Z.; Zhang, L.; Li, Y.; Yang, Y.; Yang, T.; Li, X.; Wang, K.; et al. High Detectivity Graphene-Silicon Heterojunction Photodetector. *Small* **2015**, *12*, 595–601. [CrossRef]
16. Doukas, S.; Mensz, P.; Myoung, N.; Ferrari, A.C.; Goykhman, I.; Lidorikis, E. Thermionic Graphene/Silicon Schottky Infrared Photodetectors. *Phys. Rev. B* **2022**, *105*, 115417. [CrossRef]
17. Pelella, A.; Grillo, A.; Faella, E.; Luongo, G.; Askari, M.B.; Di Bartolomeo, A. Graphene-Silicon Device for Visible and Infrared Photodetection. *ACS Appl. Mater. Interfaces* **2021**, *13*, 47895–47903. [CrossRef]
18. Ying, X.; Li, K.; Liu, L.; Wang, J.; Jiang, Y.; Xu, J.; Liu, Z. Spectral Photovoltaic Response of Graphene-Silicon Heterojunction. *Appl. Phys. Lett.* **2017**, *111*, 251106. [CrossRef]
19. Casalino, M.; Russo, R.; Russo, C.; Ciajolo, A.; Di Gennaro, E.; Iodice, M.; Coppola, G. Free-Space Schottky Graphene/Silicon Photodetectors Operating at 2 μm. *ACS Photonics* **2018**, *5*, 4577–4585. [CrossRef]
20. Selvi, H.; Unsuree, N.; Whittaker, E.; Halsall, M.P.; Hill, E.W.; Thomas, A.; Parkinson, P.; Echtermeyer, T.J. Towards Substrate Engineering of Graphene–Silicon Schottky Diode Photodetectors. *Nanoscale* **2018**, *10*, 3399–3409. [CrossRef]
21. Wong, H.; Anwar, M.A.; Dong, S. Effects of Silicon Surface Defects on the Graphene/Silicon Schottky Characteristics. *Results Phys.* **2021**, *29*, 104744. [CrossRef]
22. Jones, B.K. Electrical Noise as a Reliability Indicator in Electronic Devices and Components. *IEE Proc. Circ. Dev. Syst.* **2002**, *149*, 13–22. [CrossRef]
23. Vandamme, L.K.J. Noise as a Diagnostic Tool for Quality and Reliability of Electronic Devices. *IEEE Trans. Electron Dev.* **1994**, *41*, 2176–2187. [CrossRef]
24. Kirton, M.J.; Uren, M.J. Noise in Solid-State Microstructures: A New Perspective on Individual Defects, Interface States and Low-Frequency ($1/f$) Noise. *Adv. Phys.* **1989**, *38*, 367–468. [CrossRef]
25. Palenskis, V.; Maknys, K. Nature of Low-Frequency Noise in Homogeneous Semiconductors. *Sci. Rep.* **2015**, *5*, 18305. [CrossRef] [PubMed]
26. Balandin, A.A. Low-Frequency $1/f$ Noise in Graphene Devices. *Nat. Nanotechnol.* **2013**, *8*, 549–555. [CrossRef]
27. Arnold, H.N.; Sangwan, V.K.; Schmucker, S.W.; Cress, C.D.; Luck, K.A.; Friedman, A.L.; Robinson, J.T.; Marks, T.J.; Hersam, M.C. Reducing Flicker Noise in Chemical Vapor Deposition Graphene Field-Effect Transistors. *Appl. Phys. Lett.* **2016**, *108*, 073108. [CrossRef]
28. Kumar, A.; Kashid, R.; Ghosh, A.; Kumar, V.; Singh, R. Enhanced Thermionic Emission and Low $1/f$ Noise in Exfoliated Graphene/GaN Schottky Barrier Diode. *ACS Appl. Mater. Inter.* **2016**, *8*, 8213–8223. [CrossRef]
29. Karnatak, P.; Paul, T.; Islam, S.; Ghosh, A. $1/f$ Noise in van Der Waals Materials and Hybrids. *Adv. Phys. X* **2017**, *2*, 428–449. [CrossRef]

30. Kochat, V.; Tiwary, C.S.; Biswas, T.; Ramalingam, G.; Hsieh, K.; Chattopadhyay, K.; Raghavan, S.; Jain, M.; Ghosh, A. Magnitude and Origin of Electrical Noise at Individual Grain Boundaries in Graphene. *Nano Lett.* **2015**, *16*, 562–567. [CrossRef]
31. Karnatak, P.; Sai, T.P.; Goswami, S.; Ghatak, S.; Kaushal, S.; Ghosh, A. Current Crowding Mediated Large Contact Noise in Graphene Field-Effect Transistors. *Nat. Commun.* **2016**, *7*, 13703. [CrossRef]
32. Behera, A.K.; Harris, C.T.; Pete, D.V.; Smyth, C.M.; Muniz, M.B.; Koybasi, O.; Taniguchi, T.; Watanabe, K.; Belle, B.D.; Das, S.R. Charge-Inhomogeneity-Mediated Low-Frequency Noise in One-Dimensional Edge-Contacted Graphene Heterostructure Field Effect Transistors. *ACS Appl. Nano Mater.* **2024**, *7*, 12366–12375. [CrossRef]
33. Kakkar, S.; Karnatak, P.; Aamir, A.; Watanabe, K.; Taniguchi, T.; Ghosh, A. Optimal Architecture for Ultralow Noise Graphene Transistors at Room Temperature. *Nanoscale* **2020**, *12*, 17762–17768. [CrossRef]
34. Liu, G.; Stillman, W.; Rumyantsev, S.; Shao, Q.; Shur, M.; Balandin, A.A. Low-Frequency Electronic Noise in the Double-Gate Single-Layer Graphene Transistors. *Appl. Phys. Lett.* **2009**, *95*, 033103. [CrossRef]
35. Glemža, J.; Palenskis, V.; Gudaitis, R.; Jankauskas, Š.; Guobienė, A.; Vasiliauskas, A.; Meškinis, Š.; Pralgauskaitė, S.; Matukas, J. Low-Frequency Noise of Directly Synthesized Graphene/Si(100) Junction. *Diam. Relat. Mater.* **2022**, *127*, 109207. [CrossRef]
36. Kochat, V.; Sahoo, A.; Pal, A.N.; Eashwer, S.; Ramalingam, G.; Sampathkumar, A.; Tero, R.; Thu, T.V.; Kaushal, S.; Okada, H.; et al. Origin of $1/f$ Noise in Graphene Produced for Large-Scale Applications in Electronics. *IET Circ. Device Syst.* **2015**, *9*, 52–58. [CrossRef]
37. Haigh, S.J.; Gholinia, A.; Jalil, R.; Romani, S.; Britnell, L.; Elias, D.C.; Novoselov, K.S.; Ponomarenko, L.A.; Geim, A.K.; Gorbachev, R. Cross-Sectional Imaging of Individual Layers and Buried Interfaces of Graphene-Based Heterostructures and Superlattices. *Nat. Mater.* **2012**, *11*, 764–767. [CrossRef]
38. Chugh, S.; Mehta, R.; Lu, N.; Dios, F.D.; Kim, M.J.; Chen, Z. Comparison of Graphene Growth on Arbitrary Non-Catalytic Substrates Using Low-Temperature PECVD. *Carbon* **2015**, *93*, 393–399. [CrossRef]
39. Khan, A.; Islam, S.M.; Ahmed, S.; Kumar, R.R.; Habib, M.R.; Huang, K.; Hu, M.; Yu, X.; Yang, D. Direct CVD Growth of Graphene on Technologically Important Dielectric and Semiconducting Substrates. *Adv. Sci.* **2018**, *5*, 1800050. [CrossRef]
40. Meškinis, Š.; Gudaitis, R.; Vasiliauskas, A.; Guobienė, A.; Jankauskas, Š.; Stankevič, V.; Keršulis, S.; Stirkė, A.; Andriukonis, E.; Melo, W.; et al. Biosensor Based on Graphene Directly Grown by MW-PECVD for Detection of COVID-19 Spike (S) Protein and Its Entry Receptor ACE2. *Nanomaterials* **2023**, *13*, 2373. [CrossRef]
41. Han, Z.; Li, M.; Li, L.; Jiao, F.; Wei, Z.; Geng, D.; Hu, W. When Graphene Meets White Graphene—Recent Advances in the Construction of Graphene and *h*-BN Heterostructures. *Nanoscale* **2021**, *13*, 13174–13194. [CrossRef]
42. Wang, M.; Jang, S.K.; Jang, W.-J.; Kim, M.; Park, S.-Y.; Kim, S.-W.; Kahng, S.-J.; Choi, J.-Y.; Ruoff, R.S.; Song, Y.J.; et al. A Platform for Large-Scale Graphene Electronics—CVD Growth of Single-Layer Graphene on CVD-Grown Hexagonal Boron Nitride. *Adv. Mater.* **2013**, *25*, 2746–2752. [CrossRef]
43. Yang, W.; Chen, G.; Shi, Z.; Liu, C.-C.; Zhang, L.; Xie, G.; Cheng, M.; Wang, D.; Yang, R.; Shi, D.; et al. Epitaxial Growth of Single-Domain Graphene on Hexagonal Boron Nitride. *Nat. Mater.* **2013**, *12*, 792–797. [CrossRef]
44. Song, X.; Sun, J.; Qi, Y.; Gao, T.; Zhang, Y.; Liu, Z. Graphene/*h*-BN Heterostructures: Recent Advances in Controllable Preparation and Functional Applications. *Adv. Energy Mater.* **2016**, *6*, 1600541. [CrossRef]
45. Entani, S.; Takizawa, M.; Li, S.; Naramoto, H.; Sakai, S. Growth of Graphene on SiO_2 with Hexagonal Boron Nitride Buffer Layer. *Appl. Surf. Sci.* **2019**, *475*, 6–11. [CrossRef]
46. Won, U.Y.; Lee, B.H.; Kim, Y.R.; Kang, W.T.; Lee, I.; Kim, J.E.; Lee, Y.H.; Yu, W.J. Efficient Photovoltaic Effect in Graphene/h-BN/Silicon Heterostructure Self-Powered Photodetector. *Nano Res.* **2021**, *14*, 1967–1972. [CrossRef]
47. Stankus, V.; Vasiliauskas, A.; Guobienė, A.; Andrulevičius, M.; Meškinis, Š. Synthesis and Characterization of Boron Nitride Thin Films Deposited by High-Power Impulse Reactive Magnetron Sputtering. *Molecules* **2024**, *29*, 5247. [CrossRef] [PubMed]
48. Meškinis, Š.; Vasiliauskas, A.; Guobienė, A.; Talaikis, M.; Niaura, G.; Gudaitis, R. The Direct Growth of Planar and Vertical Graphene on Si(100) Via Microwave Plasma Chemical Vapor Deposition: Synthesis Conditions Effects. *RSC Adv.* **2022**, *12*, 18759–18772. [CrossRef] [PubMed]
49. Jankauskas, Š.; Gudaitis, R.; Vasiliauskas, A.; Guobienė, A.; Meškinis, Š. The Graphene Structure's Effects on the Current-Voltage and Photovoltaic Characteristics of Directly Synthesized Graphene/N-Si(100) Diodes. *Nanomaterials* **2022**, *12*, 1640. [CrossRef] [PubMed]
50. Hwang, J.-S.; Lin, Y.-H.; Hwang, J.-Y.; Chang, R.; Chattopadhyay, S.; Chen, C.-J.; Chen, P.; Chiang, H.-P.; Tsai, T.-R.; Chen, L.-C.; et al. Imaging Layer Number and Stacking Order through Formulating Raman Fingerprints Obtained from Hexagonal Single Crystals of Few Layer Graphene. *Nanotechnology* **2012**, *24*, 015702. [CrossRef]
51. Childres, I.; Jauregui, L.A.; Tian, J.; Chen, Y.P. Effect of Oxygen Plasma Etching on Graphene Studied Using Raman Spectroscopy and Electronic Transport Measurements. *New J. Phys.* **2011**, *13*, 025008. [CrossRef]
52. Dresselhaus, M.S.; Ado, J.; Souza, G.; Saito, R. Defect Characterization in Graphene and Carbon Nanotubes Using Raman Spectroscopy. *Philos. Trans. R. Soc. A* **2010**, *368*, 5355–5377. [CrossRef]

53. Gupta, S.; Joshi, P.; Sachan, R.; Narayan, J. Fabricating Graphene Oxide/h-BN Metal Insulator Semiconductor Diodes by Nanosecond Laser Irradiation. *Nanomaterials* **2022**, *12*, 2718. [CrossRef]
54. Parui, S.; Ruiter, R.; Zomer, P.J.; Wojtaszek, M.; van Wees, B.J.; Banerjee, T. Temperature Dependent Transport Characteristics of Graphene/N-Si Diodes. *J. Appl. Phys.* **2014**, *116*, 244505. [CrossRef]
55. Palenskis, V. The Charge Carrier Capture–Emission Process—The Main Source of the Low-Frequency Noise in Homogeneous Semiconductors. *Lith. J. Phys.* **2017**, *56*, 200–206. [CrossRef]
56. Palenskis, V.; Pralgauskaitė, S.; Matukas, J.; Glemža, J. Description of the Low-Frequency Noise for Homogeneous Semiconductors. In Proceedings of the 2023 International Conference on Noise and Fluctuations (ICNF), Grenoble, France, 17–20 October 2023. [CrossRef]
57. Luo, M.-Y.; Bosman, G.; Der, V.; Hench, L.L. Theory and Experiments of $1/f$ Noise in Schottky-Barrier Diodes Operating in the Thermionic-Emission Mode. *IEEE Trans. Electron Dev.* **1988**, *35*, 1351–1356. [CrossRef]
58. Meškinis, Š.; Gudaitis, R.; Vasiliauskas, A.; Tamulevičius, S.; Niaura, G. Multiwavelength Raman Scattering Spectroscopy Study of Graphene Synthesized on Si(100) and SiO_2 by Microwave Plasma-Enhanced Chemical Vapor Deposition. *Phys. Status Solidi Rapid Res. Lett.* **2019**, *14*, 1900462. [CrossRef]
59. Rajackaitė, E.; Peckus, D.; Gudaitis, R.; Andrulevičius, M.; Tamulevičius, T.; Volyniuk, D.; Meškinis, Š.; Tamulevičius, S. Transient Absorption Spectroscopy as a Promising Optical Tool for the Quality Evaluation of Graphene Layers Deposited by Microwave Plasma. *Surf. Coat. Technol.* **2020**, *395*, 125887. [CrossRef]
60. Angelis, C.T.; Dimitriadis, C.A.; Samaras, I.; Brini, J.; Kamarinos, G.; Gueorguiev, V.K.; Ivanov, T.E. Study of Leakage Current in N-Channel and P-Channel Polycrystalline Silicon Thin-Film Transistors by Conduction and Low Frequency Noise Measurements. *J. Appl. Phys.* **1997**, *82*, 4095–4101. [CrossRef]
61. Shin, W.; Min, K.K.; Bae, J.-H.; Yim, J.; Kwon, D.; Kim, Y.; Yu, J.; Hwang, J.; Park, B.-G.; Kwon, D.; et al. Comprehensive and Accurate Analysis of the Working Principle in Ferroelectric Tunnel Junctions Using Low-Frequency Noise Spectroscopy. *Nanoscale* **2022**, *14*, 2177–2185. [CrossRef]
62. Im, K.-S.; Shin, S.; Jang, C.-H.; Cha, H.-Y. Low-Frequency Noise Characteristics in HfO_2-Based Metal-Ferroelectric-Metal Capacitors. *Materials* **2022**, *15*, 7475. [CrossRef] [PubMed]
63. Singhal, R.; Echeverria, E.; McIlroy, D.N.; Singh, R.N. Synthesis of Hexagonal Boron Nitride Films on Silicon and Sapphire Substrates by Low-Pressure Chemical Vapor Deposition. *Thin Solid Film.* **2021**, *733*, 138812. [CrossRef]
64. Lee, J.E.; Ahn, G.; Shim, J.; Lee, Y.S.; Ryu, S. Optical Separation of Mechanical Strain from Charge Doping in Graphene. *Nat. Commun.* **2012**, *3*, 1024. [CrossRef] [PubMed]
65. Eckmann, A.; Felten, A.; Mishchenko, A.; Britnell, L.; Krupke, R.; Novoselov, K.S.; Casiraghi, C. Probing the Nature of Defects in Graphene by Raman Spectroscopy. *Nano Lett.* **2012**, *12*, 3925–3930. [CrossRef]
66. Deng, N.; Wang, P.; Zhang, E.; Guo, B.; Li, D.; Ma, B.; Fleetwood, D.M.; Tian, H.; Zhang, J. Low-Frequency $1/f$ Noise in a Graphene/Silicon X-Ray Detector. In Proceedings of the IEEE International Conference on Manipulation, Manufacturing and Measurement on the Nanoscale (3M-NANO), Xi'an, China, 2–6 August 2021; pp. 1–4. [CrossRef]
67. Wang, P.; Perini, C.; O'Hara, A.; Tuttle, B.R.; Zhang, E.X.; Gong, H.; Liang, C.; Jiang, R.; Liao, W.; Fleetwood, D.M.; et al. Radiation-Induced Charge Trapping and Low-Frequency Noise of Graphene Transistors. *IEEE Trans. Nucl. Sci.* **2018**, *65*, 156–163. [CrossRef]
68. Sze, S.M.; Ng, K.K. *Physics of Semiconductor Devices*, 3rd ed.; John Wiley & Sons, Inc.: Hoboken, NJ, USA, 2007; pp. 227–228.
69. Kajen, R.S.; Chandrasekhar, N.; Pey, K.L.; Vijila, C.; Jaiswal, M.; Saravanan, S.; Ng, A.M.H.; Wong, C.P.; Loh, K.P. Charge Transport in Lightly Reduced Graphene Oxide: A Transport Energy Perspective. *J. Appl. Phys.* **2013**, *113*, 0637110. [CrossRef]

Disclaimer/Publisher's Note: The statements, opinions and data contained in all publications are solely those of the individual author(s) and contributor(s) and not of MDPI and/or the editor(s). MDPI and/or the editor(s) disclaim responsibility for any injury to people or property resulting from any ideas, methods, instructions or products referred to in the content.

crystals

MDPI

Review

Recent Advances in the Raman Investigation of Structural and Optical Properties of Graphene and Other Two-Dimensional Materials

Grazia Giuseppina Politano [1,*] and Carlo Versace [2,3]

[1] Department of Information Engineering, Infrastructures and Sustainable Energy (DIIES), University "Mediterranea" of Reggio Calabria, 89122 Reggio Calabria, Italy
[2] Dipartimento di Fisica, Università della Calabria, 87036 Rende, Italy; carlo.versace@fis.unical.it
[3] Licryl CNR/Nanotec c/o Dipartimento di Fisica, Università della Calabria, 87036 Rende, Italy
* Correspondence: grazia.politano@unirc.it

Abstract: This research work presents our recent advancements in the study of graphene and other two-dimensional (2D) materials. This review covers studies on graphene oxide (GO) thin films deposited on various substrates, including titanium and silver thin films. The analysis reveals the role of surface-enhanced Raman scattering (SERS) and the influence of film thickness and laser intensity on the observed Raman peaks. Investigations of a silver/GO/gold sandwich structure demonstrate the presence of sharp Raman modes attributed to localized surface plasmon resonances. This review also discusses the characterization of exfoliation-synthesized graphene nanoplatelet (GNPs) thin films and of single-layer graphene deposited via chemical vapor deposition (CVD). The optical properties of molybdenum disulfide (MoS_2) films are also explored. The utilization of micro-Raman spectroscopy provides valuable insights into the structural and optical properties of graphene and other 2D materials. These results hold the potential to drive advancements in various applications, such as electronics, photonics, and nanocomposites.

Keywords: graphene; graphene oxide; micro-Raman; molybdenum disulfide; two-dimensional materials

1. Introduction

Graphene and other two-dimensional (2D) materials display exceptional characteristics, making them highly advantageous for a diverse range of applications across various fields [1–8], including flexible nanoelectronics, sensors, and photodetectors [9–17].

Graphene exhibits special physical and electrical properties. It demonstrates broad light absorption, a linear dispersion band structure, significant charge-carrier mobility, and remarkable optical qualities. For instance, it offers high transparency within the visible spectrum [18], photo-response capabilities extending into the Terahertz frequency range, and customizable infrared optical absorbance [19]. Furthermore, its unique band structure and electron transport properties make it a promising candidate for developing high-performance modulators, sources, and detectors designed for operation in the infrared (IR) segment of the electromagnetic spectrum [20]. The optical characteristics of graphene films vary significantly depending on the fabrication method used, whether it is exfoliation or chemical vapor deposition (CVD) [14]. CVD, a method intended for producing high-quality few-layer graphene, has been successfully applied in ultradense photonic, optoelectronic, and electronic devices, as described in previous research [21,22].

Recently, there has been growing interest in combining the unique properties of graphene-based materials with silver nanostructures, aiming for their utilization in the fields of plasmonics and metamaterials applications [23–25]. Silver nanoparticles and nanostructures, with localized surface plasmonic resonance spanning from the visible to the near-infrared range, find utility in various applications, particularly in surface-enhanced Raman scattering (SERS) [26]. In this context, materials based on graphene, for instance, graphene oxide (GO) and reduced graphene oxide (RGO), have demonstrated their potential as SERS substrates by generating strong chemical enhancement [27].

Nevertheless, even with substantial endeavors by the scientific community, attaining widespread accessibility to graphene samples remains a considerable obstacle, hindering the exploitation of its exceptional application capabilities [28]. Consequently, while single-layer graphene is not readily accessible for large-scale bulk applications, materials related to graphene, including RGO, few-layer graphene, multilayer graphene, and graphene nanoplatelets (GNPs), have emerged as pioneering materials with numerous potential applications [29].

GO has emerged as a novel nanomaterial based on carbon that provides a different pathway to graphene [30]. By utilizing robust oxidizing agents in processes like the Hummers method, graphite undergoes oxidation, leading to the incorporation of diverse oxidized functional groups, such as epoxide, carbonyl, carboxyl, and hydroxyl groups, into its molecular framework [31–33]. This process widens the gaps between layers and imparts hydrophilic properties to the material. As a result, graphite oxide can be effectively dispersed in water via ultrasonic treatment, ultimately yielding the fabrication of individual or a limited number of GO layers [34].

GNPs, which are another graphene-based material, possess several advantageous characteristics akin to those of monolayer graphene [35,36]. When comprised of a monolayer to a small number of layers, these sheets consist of sp^2-bonded carbon atoms that interlock, forming particles with a 2D nanoarchitecture on the scale of nanometers. These particles are realized through the exfoliation of low-cost graphite flakes, followed by chemical treatment and the reduction of graphite oxide nanoplatelets [37–39].

Recent advancements have also been reported in the field of metal dichalcogenides (WS_2, MoS_2, $MoSe_2$, WSe_2) films [40]. Transition metal dichalcogenide (TMD) materials, because of their exceptional optoelectronic characteristics, hold various potential applications when combined with conducting materials [41]. Some of these applications include hydrogen evolution, Li-ion batteries, solar cells, and supercapacitors [42,43].

In this review, we offer a comprehensive analysis of the recent advancements in the optical and structural properties of graphene and other 2D materials based on the authors' research [44–49].

2. Structure of Graphene, Graphene Oxide, and Reduced Graphene Oxide

Graphene consists of a single layer of carbon atoms arranged in a hexagonal lattice and originates from graphite [50]. Its unique properties include sp^2 hybridization and an incredibly slim atomic thickness measuring 0.345 nm [51]. A perfect graphene sheet has a flat structure with minor ripples, and its carbon atoms form a trigonal pattern [52]. Various methods have been developed to fabricate graphene, including CVD [53], epitaxial growth [54], micromechanical exfoliation [55], and epitaxial growth on insulating surfaces like SiC [56]. These techniques have significantly advanced research on monolayer graphene.

GO, on the other hand, is a derivative of graphene that contains oxygen-functional groups like epoxide, carbonyl, carboxyl, and hydroxy [57]. These groups render GO sheets water-soluble; their removal leads to flocculation and precipitation. GO features a complex structure with oxygen-rich domains interspersed among areas where carbon atoms retain their sp^2 hybridization [58].

Graphite and graphite oxide, although related, have distinct properties [59]. Graphite oxide is created by oxidizing graphite, which introduces oxygen functionalities that increase the distance between layers and make the material water-dispersible [60]. Exfoliation of graphite oxide through sonication in water results in the formation of GO [61]. The primary difference between graphite oxide and GO is in the number of layers: graphite oxide is multilayered, while GO dispersions mainly consist of monolayer flakes, with some multilayer flakes present [62].

Zeta potential measurements show that GO sheets in water carry negative charges [63], which leads to electrostatic repulsion among the sheets and results in a stable aqueous suspension. The utilization of organic compounds in chemically altering GO sheets can lead to the attainment of uniform dispersions in organic solvents [64].

A widely employed technique for synthesizing GO is the adapted Hummers method, which relies on oxidizing graphite by using a blend of concentrated sulfuric acid, sodium nitrate, and potassium permanganate [65]. While the exact chemical arrangement of GO remains not entirely comprehended, it is considered partially amorphous, with proposed structural models consisting of hydroxyl and epoxy functional groups [66]. Analysis conducted using atomic force microscopy has unveiled that the thickness of a single-layer GO sheet can exhibit fluctuations ranging from 0.7 to 1.4 nm, notably surpassing the thickness of an ideal graphene monolayer [67].

Various techniques, such as drop-casting [68], dip-coating [69], spraying [70], spin-coating [71], and electrophoresis [72], can be used to deposit GO sheets as thin films on different substrates.

GO and graphene have also been extensively used to obtain composite coatings, enhancing the wear behavior and other mechanical properties of the substrate. For example, in Ref. [73], an enhanced wear behavior in nickel/graphene composite coatings is reported. Similarly, in Ref. [74], the properties of nickel-based composite coatings modified with GO are investigated.

The electrical properties of GO sheets differ significantly from those of pristine graphene. As a consequence, films composed of GO tend to manifest resistance values that commonly exceed 10^{12} Ω/sq [75].

A partial reduction of GO can yield sheets resembling graphene by eliminating oxygen-related groups [76]. Chemical reduction methods typically use reducing agents such as hydrazine or involve creating highly alkaline conditions [77]. Ascorbic acid has emerged as a promising alternative to hydrazine [78]. The thermal reduction process for GO is commonly conducted at temperatures exceeding 200 °C within inert or reducing surroundings, with an enhanced efficacy observed at elevated temperatures [79].

Understanding the methods of reducing GO is pivotal, as the process not only alters its hydrophilicity but also has a profound impact on its electrical conductivity, thereby influencing its suitability for various applications.

These reduced GO sheets exhibit reduced hydrophilicity and display a propensity to aggregate within the solution, stemming from the absence of oxygenated functionalities [80]. The chemical reduction of GO partially restores its conductivity [81], yet it remains significantly inferior by several orders of magnitude when compared to the pristine graphene counterpart. The degree of this reduction determines the conductivity of the resulting RGO, which can range from 0.05 to 500 S/cm and is intricately linked to the proportion of graphitic (sp^2) to oxidized (sp^3) zones [82].

It remains a significant challenge to develop a production method for RGO that closely mimics the properties of pristine graphene. However, due to its moderate conductivity and the ability to be produced into a thin film through an aqueous solution, RGO is attractive for use in electronic devices. Additionally, GO and RGO find applications in nanocomposite materials, polymer composites, energy storage, and biomedical uses like drug delivery and catalysis [83].

3. State of the Art

Raman spectroscopy has become an indispensable tool for comprehensively characterizing graphene and its derivatives like GO and RGO [84,85]. The technique's ability to identify characteristic vibrational modes in materials offers an understanding of the structural and chemical compositions of these graphene-based materials [86]. For example, Neeraj Sharma and colleagues [87] showed that the presence or absence of functional groups significantly alters the material's vibrational modes, evident from changes in the Raman spectra [86]. This is particularly valuable for researchers aiming to tailor the properties of graphene-based materials for specific applications, such as optoelectronics or sensing.

The G and D peaks in the Raman spectra serve as critical markers for the structural integrity and chemical modifications of graphene-based materials [84]. The G peak, a fundamental graphitic peak, is ubiquitous in all hexagonal graphitic structures, while the D peak arises due to imperfections and impurities introduced by functional groups [84].

Jiang-Bin Wu and colleagues [88] further elaborate on how Raman spectroscopy can study the impact of various external factors like doping agents, mechanical forces, and environmental conditions on single-layer graphene. Additionally, the authors provide insights into the identification of material defects and layer counts through Raman spectral features.

K. Kanishka H. De Silva et al. [89] further extend this discussion by emphasizing the challenges posed by the abundant lattice defects in GO and RGO. Their work provides new insights into the optical visibility and Raman signals of these materials when deposited on different Si-based substrates. They found that dielectric substrates with lower reflectance in the visible range offer better optical contrast and more intense Raman signals. This advancement is significant because it implies that substrate choice can affect the quality of Raman spectroscopic data and, consequently, our understanding of the material's structural and optical properties. Moreover, De Silva et al. [89] introduced the importance of the D′ band in the Raman spectra for a more comprehensive understanding of the structural variations in monolayer flakes of GO and RGO.

The existing literature on Raman spectroscopy for characterizing graphene and 2D materials is extensive [90]; however, our review introduces several unique contributions that set it apart from previous works.

A standout feature of our review explores magnetron sputtering as a deposition technique and its implications for SERS. While SERS is a well-known technique for enhancing Raman signals, its application in 2D materials deposited via magnetron sputtering is less commonly discussed. Our review delves into how magnetron sputtering can influence the SERS properties of 2D materials, offering new avenues for highly sensitive detection methods in both industrial and research settings.

Another unique aspect of our review is the focus on solution-based MoS_2. While MoS_2 has been extensively studied, few works have addressed the unique challenges and opportunities presented by solution-based methods. Our review fills this gap by discussing the Raman characterization of solution-based MoS_2, offering a more comprehensive understanding of this material in liquid-phase environments. This is particularly relevant for applications where solution-based methods offer advantages in terms of scalability and cost-effectiveness.

4. Materials and Methods

Micro-Raman spectra [44–49] were obtained using a Horiba-Jobin Yvon micro-probe instrument with an approximate spectral resolution of 2 cm^{-1}. This setup incorporated a CCD detector boasting dimensions of 256 × 1024 pixels, cooled to −70 °C, and was coupled with a diode laser emitting a 532 nm line, generating an output power of 50 mW. A 50× Mplan Olympus objective was used to focus the laser, yielding a laser spot with an apparent diameter of approximately 2–3 μm. To avert the structural modifications stemming from laser-induced heating, an OD2 filter with a 1% transmission rate was deployed.

The GO dispersions of the GO discussed were sourced from Sigma Aldrich (St. Louis, MO, USA) and Punto Quantico (Roma, Italy). The dispersion of MoS_2 was bought from Sigma Aldrich (St. Louis, MO, USA). Monolayer graphene samples, grown via CVD, were procured from Graphenea Co. (San Sebastián, Spain).

5. Results and Discussion

This section provides a comprehensive analysis of graphene and other 2D materials, employing micro-Raman spectroscopy as the principal investigative tool. The focus spans from the electrophoretic deposition techniques of GO to the complex Ag/GO/Au sandwich structure. Each subsection aims to elucidate the unique structural and spectral characteristics of these materials, thereby contributing to the understanding of their potential applications and underlying mechanisms.

5.1. Thin Films of Graphene Oxide and Reduced Graphene Oxide Deposited by Electrophoresis

The electrophoretic deposition (EPD) method is highly adaptable for creating GO-based thin films. Charged GO platelets in a suspension move toward an electrode with an opposing charge under an applied electric field, forming a compact film. EPD offers advantages like a rapid deposition rate, controllable thickness, uniformity, and scalability. It requires electrically conductive substrates like Cu, Ni, Al, stainless steel, and Ti [91].

Figure 1a presents the Raman spectrum of electrophoretically deposited GO on a titanium layer before the annealing treatment.

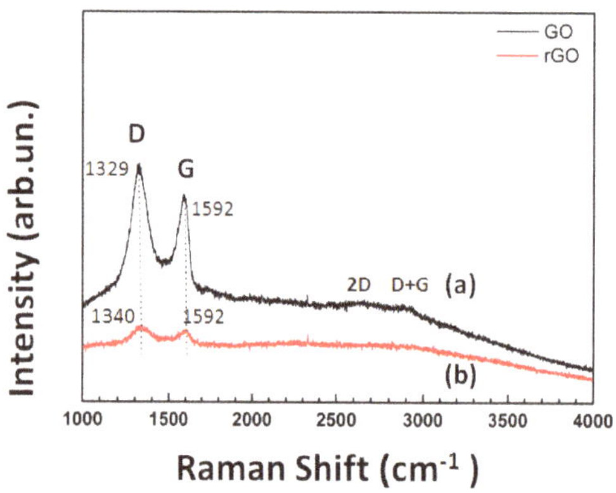

Figure 1. Raman spectra of electrophoretically deposited graphene oxide on a titanium layer (**a**) prior to annealing and (**b**) after annealing at 400 °C. This content has been reproduced from reference [44] with permission granted by AIP Publishing. Copyright 2016, AIP Publishing.

The Raman spectrum shows characteristic peaks, specifically at 1329 cm^{-1} for the D band and 1592 cm^{-1} for the G band of GO. Additionally, the 2D band ($\approx$2640 cm^{-1}) and the second-order band, D+G ($\approx$2912 cm^{-1}), can be observed. These spectral features suggest a considerable level of disorder in the GO.

Figure 1b illustrates the GO film deposited on a titanium layer after the annealing at 400 °C. In this case, the D band is detected at about 1343 cm^{-1}, indicating either a different level of carbon layer oxidation or varying mechanical stresses on the graphene layer.

5.2. Graphene Oxide on Magnetron-Sputtered Silver Thin Films for Surface-Enhanced Raman Spectroscopy Applications

GO's potential in SERS is increasingly recognized. In this section, we look at how GO behaves when deposited on silver substrates deposited through magnetron sputtering.

Visual examination reveals two distinct types of GO deposition on the magnetron-sputtered silver substrate, as evident in Figure 2's optical microscopy image.

The first type maintains optical properties like color and reflectivity, similar to the unaltered silver surface. These "bright" areas indicate the presence of a single or a few layers. Conversely, the remaining areas, referred to as "dark", appear brown and show reduced reflectivity compared to the original silver surface.

The spectrum in Figure 3b comes from a "medium dark zone", where GO is dip-coated on a 10 nm magnetron-sputtered silver substrate using low laser intensity (OD2). It reveals that the D and G Raman bands align with the frequencies observed in Figure 3a, where GO was dip-coated on untreated glass using the same experimental parameters. Significantly, the intensities of the D and G bands are markedly increased.

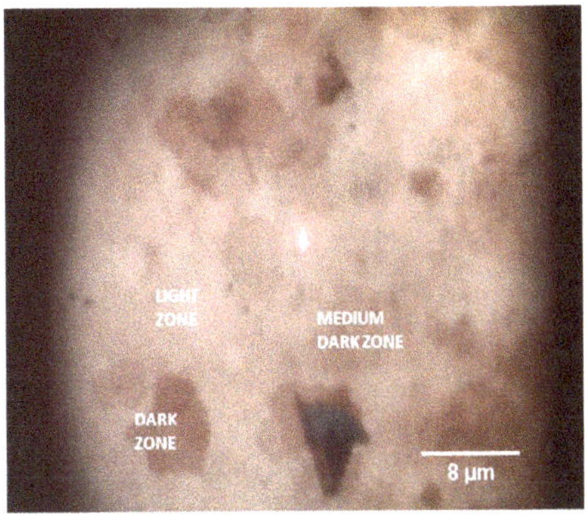

Figure 2. Optical image of a representative area showing dip-coated graphene oxide films on a 10 nm magnetron-sputtered silver substrate. The laser highlights a "light" area. This content has been reproduced from reference Ref. [45] with permission granted by Elsevier. Copyright 2018, Elsevier.

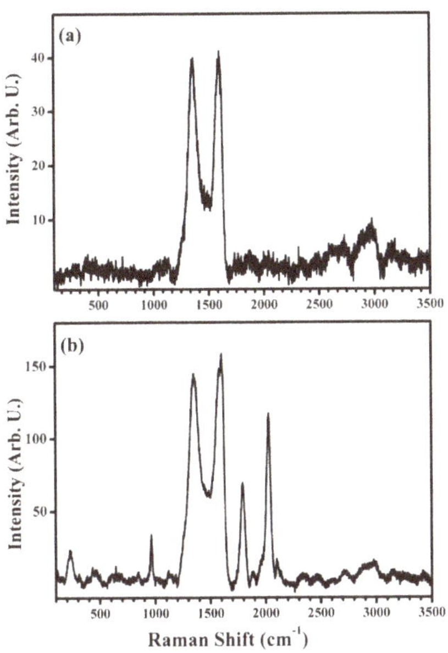

Figure 3. Raman spectra of graphene oxide films dip-coated on bare glass at low laser intensity (OD2) (**a**) and graphene oxide films dip-coated on a 10 nm magnetron-sputtered silver/glass substrate at low laser intensity (OD2) (**b**). This content has been reproduced from reference Ref. [45] with permission granted by Elsevier. Copyright 2018, Elsevier.

In Figure 3b, we observe distinct peaks at 235 cm^{-1}, 967 cm^{-1}, 1793 cm^{-1}, and 2033 cm^{-1}. These distinctive peaks were mostly pronounced in the "medium dark zones" compared to the "bright zones" and "dark zones." Specifically, the peaks at 1793 cm^{-1} and 2033 cm^{-1} can be attributed to oxygen functionalities on the GO surface, likely C=O stretching vibrations from the carbonyl and carboxylic groups. The peaks at 235 cm^{-1} and 967 cm^{-1} can be associated with the existence of oxide in the silver thin films. This interaction between oxygen and silver thin films could result from the subsequent reactions occurring after deposition in ambient conditions.

Figure 4 presents the Raman spectrum from the same spot in the "medium dark zone", as indicated in Figure 3b. However, this measurement uses the full laser intensity without any filtering.

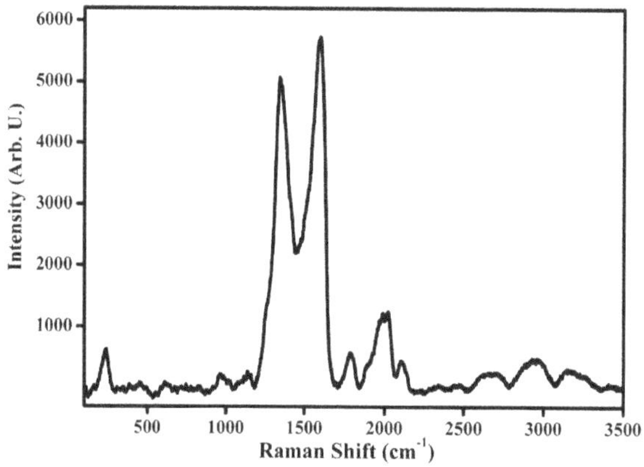

Figure 4. Raman spectra of graphene oxide films fabricated via dip-coating on a 10 nm magnetron-sputtered silver/glass substrate under unfiltered full laser intensity. This content has been reproduced from Ref. [45] with permission granted by Elsevier. Copyright 2018, Elsevier.

Figure 4 shows a shift to lower frequencies in the D band, around 1343 cm^{-1}, compared to Figure 3b. This could indicate a different level of oxidation in the GO layer. Meanwhile, the G band remained at the identical frequency of 1595 cm^{-1}, as shown in Figure 3b. Moreover, a prevalent reduction in the intensity of the features at 235 cm^{-1}, 967 cm^{-1}, 1793 cm^{-1}, and 2033 cm^{-1} is evident when compared with the spectrum presented in Figure 3b. These observations suggest a significant variation in the coupling between GO and silver when subjected to full laser power irradiation, leading to irreversible damage to the GO layer.

5.3. Micro-Raman Spectroscopy Measurements of Spin-Coated MoS_2 Films

In this subsection, we explore the properties of spin-coated MoS_2 films on Si/SiO_2 substrates using micro-Raman spectroscopy.

The primary Raman modes of MoS_2 include E_{1g} (286 cm^{-1}), $E_{2g}{}^1$ (383 cm^{-1}), A_{1g} (408 cm^{-1}), and $E_{2g}{}^2$ (32 cm^{-1}). The E_{1g}, $E_{2g}{}^1$, and $E_{2g}{}^2$ modes are indicative of in-plane Raman activity, whereas the A_{1g} mode signifies out-of-plane vibrations. Figure 5 shows typical Raman spectra from MoS_2 films spin-coated onto Si/SiO_2 substrates. These spectra originate from distinct locations on the samples.

In Figure 5a, only the $E_{2g}{}^1$ and A_{1g} modes are present, appearing at approximately 380 cm^{-1} and 405 cm^{-1}, correspondingly.

The location of the higher-frequency mode signifies the presence of a monolayer MoS_2 specimen, whereas the alternate mode implies a structure composed of multiple layers. These results strongly suggest that the content depicted in Figure 5a pertains to 2H-MoS_2.

In Figure 5b, besides the features seen in Figure 5a, distinct peaks around 290 cm^{-1} and 299 cm^{-1} are clearly visible. While the E_{1g} mode is typically disallowed in backscattering geometry for Raman scattering, these two modes have been designated as E_{1g}. Specifically, the mode at 299 cm^{-1} might be linked to 1T-MoS_2, while the signal at 290 cm^{-1} could be ascribed to the amorphous phase of MoS_2. The capacity to detect the E_{1g} mode, even in a backscattering arrangement, is attributed to the inherent disorder in the amorphous phase.

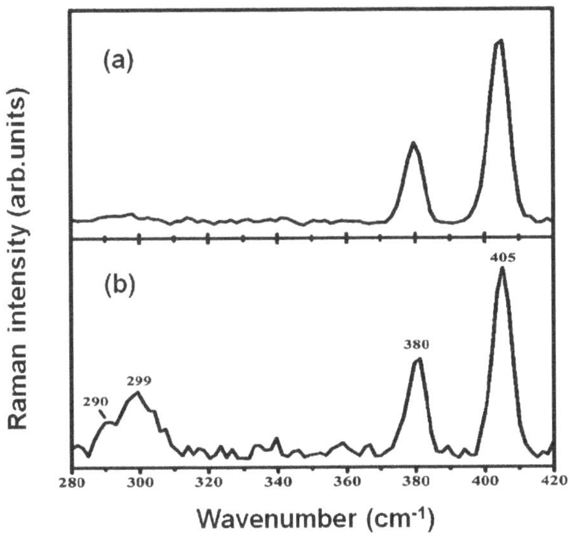

Figure 5. Representative micro-Raman spectra collected from MoS_2 films spin-coated onto Si/SiO_2 substrates (**a**,**b**). This content has been reproduced from Ref. [48] with permission granted by Elsevier. Copyright 2021, Elsevier.

5.4. Micro-Raman Spectroscopy Measurements of Graphene Nanoplatelets

GNPs are increasingly recognized for their unique properties in the field of graphene-based materials. Herein, we examine thin films of GNPs on Si/SiO_2 substrates to understand their structural features.

GNPs exhibit characteristic Raman bands, including the G band (resulting from phonons at the Brillouin zone center) and the D band (associated with amorphous regions), as well as the overtone 2D band and smaller bands originating from the second/third order modes. Analyzing these Raman bands provides insights into various properties of graphene systems, such as thickness, lattice order, mechanical strain, and chemical doping.

Visual examination of GNP thin films on Si/SiO_2 substrates, as shown in Figure 6 (panels a–h), reveals a mostly uniform surface with some thicker areas called "islands".

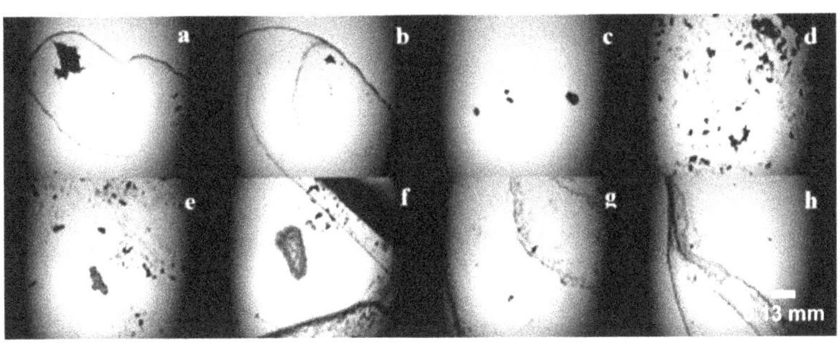

Figure 6. Optical images depicting thin films of graphene nanoplatelets on Si/SiO_2 (**a**–**h**). This content has been reproduced from Ref. [47] with permission granted by IOP Publishing. Copyright 2019, IOP Publishing.

A representative Raman spectrum obtained from these "islands" is illustrated in Figure 7.

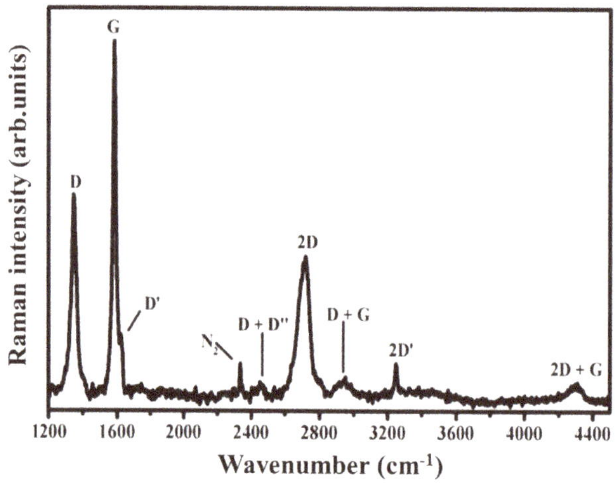

Figure 7. Representative Raman spectrum acquired from the islands of graphene nanoplatelet thin films on Si/SiO_2. This content has been reproduced from Ref. [47] with permission granted by IOP Publishing. Copyright 2019, IOP Publishing.

In addition to the silica substrate signals at 518 cm^{-1} and 964 cm^{-1}, which are not shown in Figure 7, unique characteristics associated with the GNPs become evident. Specifically, the D, G, and 2D bands appear at 1346 cm^{-1}, 1582 cm^{-1}, and 2719 cm^{-1}, correspondingly. Situated around 1620 cm^{-1} is the D′ band. Furthermore, combination bands like D + D″ (~2462 cm^{-1}), D + G (~2951 cm^{-1}), 2D + G (4300 cm^{-1}), and 2D′ (3249 cm^{-1}) can be identified. These spectral features indicate the existence of irregular graphene structures.

5.5. Micro-Raman Spectroscopy Measurements of CVD-Grown Monolayer Graphene

CVD is commonly employed as the preferred method for synthesizing high-quality monolayer graphene. This subsection delves into how micro-Raman spectroscopy serves as a powerful tool for characterizing this type of graphene.

In Figure 8, we present two typical micrographs. Figure 8a exhibits a central segment of the specimen, where the relatively darker portions are marked as "A", while a more luminous stripe is identified as "C".

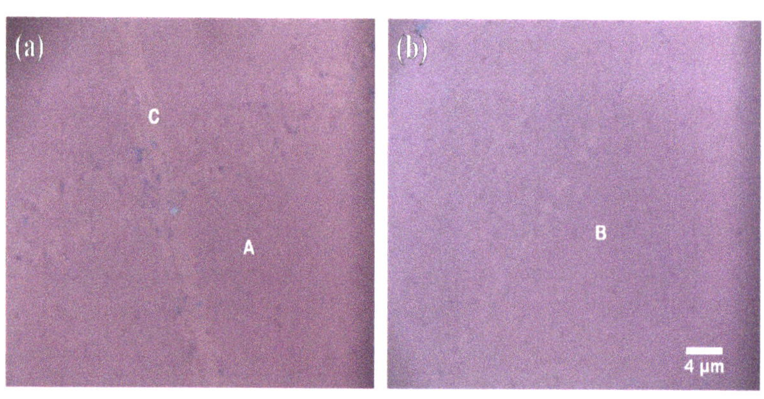

Figure 8. Micrographs illustrating the central zone (A zone) and the border zone (B zone) of the sample. Darker regions and a brighter stripe are observed in the central zone (**a**), while a homogeneous region is visible in the border zone (**b**). This content has been reproduced from Ref. [49] with permission granted by Elsevier. Copyright 2019, Elsevier.

Figure 8b highlights an edge section of the specimen, designating a uniform area as "B". Spectral features collected from zone A are shown in Figure 9, covering the wavenumber range of 400 cm^{-1} to 1100 cm^{-1} (a) and 1200 cm^{-1} to 3400 cm^{-1} (b).

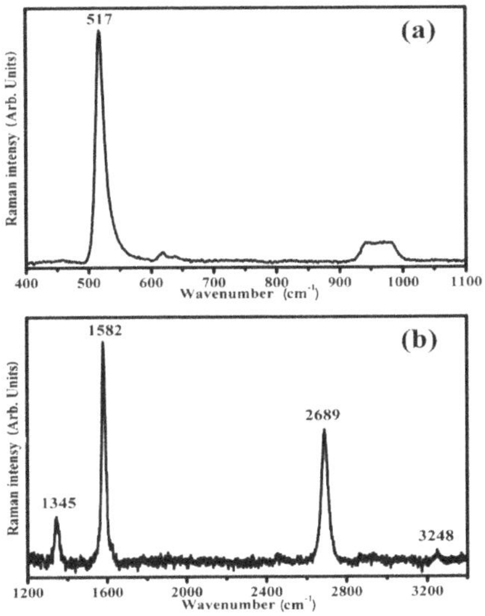

Figure 9. Illustrative Raman spectra from the central zone (designed as zone A) of the graphene/Si samples. Spectra range from 400 cm^{-1} to 1100 cm^{-1} (**a**) and 1200 cm^{-1} to 3400 cm^{-1} (**b**). This content has been reproduced from Ref. [49] with permission granted by Elsevier. Copyright 2019, Elsevier.

The prominent Raman features in these spectra occur at 1345 cm^{-1}, 1582 cm^{-1}, 2689 cm^{-1}, and 3248 cm^{-1}. The minor band at 1345 cm^{-1} corresponds to the graphene D band, while the feature at 1582 cm^{-1} represents the G band. The signals detected at 2689 cm^{-1} and 3248 cm^{-1} correspond to the overtones 2D and 2D′, respectively.

Raman spectra obtained from "B" are depicted in Figure 10, spanning wavenumber ranges from 400 cm^{-1} to 1100 cm^{-1} (a) and from 1200 cm^{-1} to 3400 cm^{-1} (b).

In these spectra, the D band at 1351 cm^{-1} and the G band at 1589 cm^{-1} show slight blue shifts, while the 2D band at 2689 cm^{-1} and the 2D′ band at 3248 cm^{-1} remain at the same frequencies as in Figure 9. Furthermore, within the spectrum of zone B, a discernible band at 2454 cm^{-1} emerges, labeled as D″ + D′, with D″ being apparent at around 1073 cm^{-1}.

Figure 11 displays the Raman spectra from region C, covering wavenumbers from 400 cm^{-1} to 1100 cm^{-1} (a) and 1200 cm^{-1} to 3400 cm^{-1} (b).

In this case, a significant blue shift is observed in the G band at 1595 cm^{-1}, while the D band at 1346 cm^{-1}, the 2D band at 2681 cm^{-1}, and the 2D′ band at 3242 cm^{-1} all show downward shifts. Additionally, within the spectrum of region C, the presence of the D″ + D′ band becomes evident.

The ratio between the intensity of the 2D band and the G band (I_{2D}/I_G) provides information about the amount of graphene layers. An I_{2D}/I_G value exceeding 2 signifies single-layer graphene, a value between 1 and 2 suggests bilayer graphene, and a value lower than 1 signifies the presence of three or more layers.

For the samples being studied, the I_{2D}/I_G ratios in zones A and B, as shown in Figures 9 and 10, are 1.04 and 1.68, respectively. In zone C (Figure 11), the I_{2D}/I_G ratio value stands at 2.63. Consequently, while single-layer graphene predominantly prevails, there are sections displaying bilayer graphene arrangements in both the central and border zones. In contrast, single-layer graphene is discernible in the brighter stripes.

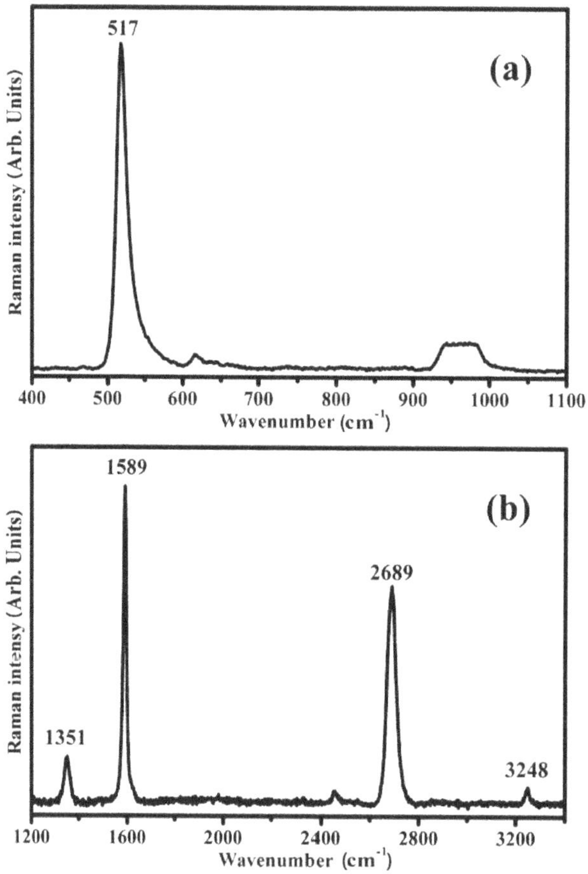

Figure 10. Illustrative Raman spectra acquired from the border region (zone B) of the graphene/Si samples. Raman spectral profiles spanning the range of 400 cm^{-1} to 1100 cm^{-1} (**a**) and 1200 cm^{-1} to 3400 cm^{-1} (**b**). This content has been reproduced from Ref. [49] with permission granted by Elsevier. Copyright 2019, Elsevier.

5.6. Micro-Raman Spectroscopy Measurements of Ag/GO/Au Sandwich Structure

The Ag/GO/Au sandwich structure is an interesting configuration that combines silver, GO, and gold.

For reference, Figure 12a presents the illustrative Raman spectrum collected from a thin GO film (8 nm) dip-coated onto an untreated glass substrate. This measurement was conducted using low laser intensity (OD2).

The characteristic traits of GO, represented by the D and G bands, become noticeable at approximately 1365 cm^{-1} and 1595 cm^{-1}, respectively. On the contrary, the spectrum derived from the Ag/GO/Au sandwich structure displays a markedly different profile, featuring several pronounced peaks that exceed the expected intensity levels of the GO bands previously described in the reference spectrum. This unique spectral behavior is often observed in areas of the sample that appear darker under microscopic examination, generally indicative of the presence of multiple GO layers.

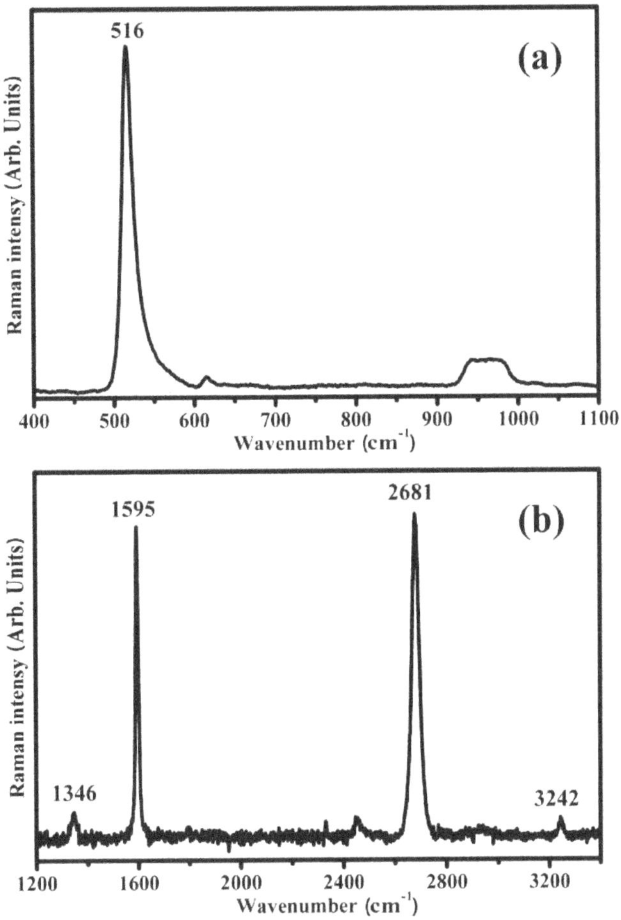

Figure 11. Exemplary Raman spectra obtained from imperfections within the graphene layer (zone C) of the graphene/Si samples. Raman spectral profiles spanning the range of 400 cm^{-1} to 1100 cm^{-1} (**a**) and 1200 cm^{-1} to 3400 cm^{-1} (**b**). This content has been reproduced from Ref. [49] with permission granted by Elsevier. Copyright 2019, Elsevier.

Figure 12b illustrates a Raman spectrum acquired from a "dark" region within the Ag/GO/Au sandwich structure. A low laser intensity setting (OD2) was used to highlight this particular spectral behavior. Although the D and G bands typical of GO are clearly visible at approximately 1365 cm^{-1} and 1595 cm^{-1}, similar to the reference spectrum in Figure 12a, the most notable observation is the appearance of multiple additional peaks with sharp definitions. In some cases, these peaks are strikingly prominent, displaying an even higher intensity than the GO bands.

The most noticeable peaks are found at 1077 cm^{-1}, 1122 cm^{-1}, 1240 cm^{-1}, 1286 cm^{-1}, 1478 cm^{-1}, and 1976 cm^{-1}. These distinctly sharp peaks are linked to localized vibrational modes of molecular groups that attach themselves to the GO network and establish a close interaction with metal particles or protrusions, commonly referred to as "hot spots."

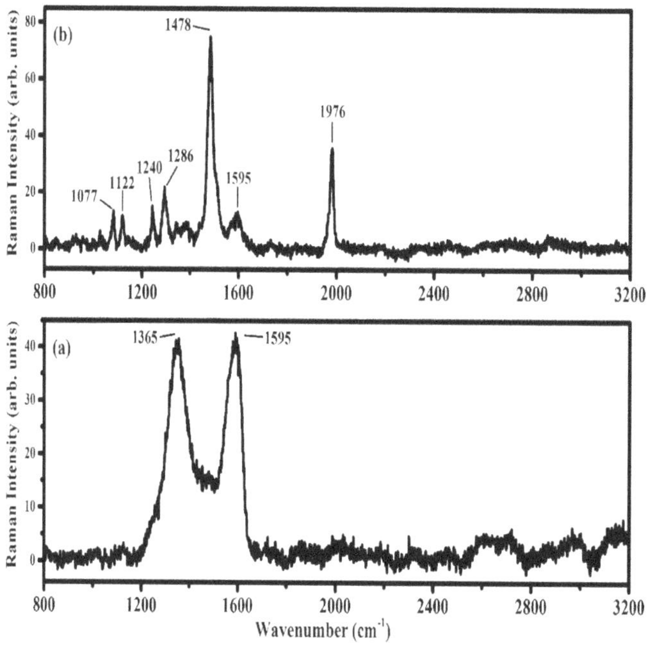

Figure 12. Representative Raman spectra taken under low laser intensity (OD2 filter) conditions, showcasing graphene oxide thin film dip-coated onto an untreated glass sample (**a**) and an Ag/graphene oxide/Au sandwich structure (**b**). This content has been reproduced from Ref. [46] with permission granted by IOP Publishing. Copyright 2019, IOP Publishing.

These supplementary Raman modes exhibit instability under heat treatment. Specifically, under high laser intensities, the pronounced peaks associated with the "hot spots" disappear, making the resulting Raman spectrum closely align with that shown in Figure 12a. This change mainly shows the disordered features of the GO network. The disappearance of these peaks can be understood as the result of either photolysis or thermolysis affecting the molecular groups that were amplifying the Raman signals at the "hot spots". Despite the higher laser power, the fundamental carbon structure of GO remains unaffected.

6. Conclusions, Applications, and Outlook

6.1. Conclusions

This review outlines our recent research on graphene and other 2D materials, focusing on the use of micro-Raman spectroscopy for characterization. Among the key findings:

- GO Thin Films on Ti/Glass Substrates: We used micro-Raman spectroscopy to study the properties of GO layers that were deposited on titanium substrates through electrophoretic deposition. This technique proved invaluable for gaining insights into the structural and optical characteristics of the GO layers.
- GO Thin Films on Magnetron-Sputtered Silver: A significant portion of the review is dedicated to the study of GO thin films fabricated by dip-coating onto magnetron-sputtered silver substrates. Micro-Raman analysis reveals that these films exhibit SERS activity.
- Optical Properties of MoS_2: This review also includes an exploration of the optical properties of solution-based MoS_2, with micro-Raman spectroscopy employed for its characterization.
- Investigation of GNPs: This review includes a section on the investigation of GNPs.
- Monolayer Graphene on SiO_2/Si Substrates: The optical properties of monolayer graphene samples grown via CVD on SiO_2/Si substrates are discussed.

- Silver/GO/Gold Sandwich Structure: The review reports on the micro-Raman investigation of a silver/GO/gold sandwich structure. Notably, sharp Raman modes have been observed, sometimes even exhibiting greater intensity than typical GO bands, particularly under low laser power conditions.

6.2. Applications of Graphene and Other Two-Dimensional Materials

Raman spectroscopy is a critical tool for characterizing the properties of graphene, especially in the development of graphene-based field-effect transistors (FETs). By identifying the doping levels, Raman spectroscopy enables the fine-tuning of the electronic properties of these transistors, directly affecting their speed and energy efficiency. Additionally, the technique provides valuable data on heat generation and dissipation within the FETs, contributing to their long-term stability and performance [92–95].

The technique also plays a significant role in advancing energy storage devices. Specifically, it helps researchers understand the distortion of electrode material during charge and discharge processes. This understanding is vital for enhancing both the efficiency and longevity of batteries and other energy storage systems based on graphene [96].

In optoelectronic applications like solar cells and organic light-emitting diodes (OLEDs) [97], graphene's high optical transparency, electrical conductivity, and mechanical flexibility make it a material of choice. Raman spectroscopy aids in the identification of interlayer coupling and the number of graphene layers in these devices, which is crucial for optimizing their performance.

Moreover, in nanoelectromechanical systems (NEMS), the technique is used to probe the local stress within multilayer graphene cantilevers [98].

Raman spectroscopy is also instrumental in the study of van der Waals 2D heterostructure devices, such as those combining graphene and MoS_2 [99]. The method allows for the in situ examination of charge transfer within these heterostructures, providing essential insights into the electronic interplay between the various layers.

While the applications of Raman spectroscopy in areas like FETs, energy storage, and optoelectronics are well-established, its potential in sensing technologies is gaining increasing attention [100].

It is worth noting that MoS_2, a prominent member of the transition metal dichalcogenides family, has demonstrated significant applicability across various technological domains. Notably, MoS_2 has been studied for its role as a spin-valley filter, as outlined in [101]. Additionally, its intrinsic optoelectronic characteristics make it suitable for phototransistor devices, as detailed in [102]. MoS_2 has also been explored for its potential in rectifying contacts, as reported in [103], and in tunneling field-effect transistors, as described in [104].

The study of transition metal dichalcogenide monolayers (MX_2), including MoS_2, has revealed the importance of metal-induced gap states (MIGS) in spin-valley filters [101]. These MIGS are predicted to mediate valley- and spin-resolved charge transport near the ideal electrode/MX_2 interface, thus initiating filtering. This insight is particularly crucial when the channel length is diminished as MIGS begin to govern the overall valley-spin transport in the tunneling regime. The findings offer design guidelines for efficient valley-spin filter devices, which could revolutionize information processing technologies.

In addition to its role in valley-spin filtering, MoS_2 has been extensively studied for its intrinsic optoelectronic characteristics [102]. Raman spectroscopy serves as a vital tool for understanding these properties, especially when MoS_2 is integrated into functional devices like phototransistors. Utilizing a fully transparent van der Waals heterostructure, researchers have been able to reveal the intrinsic photoresponsivity characteristics of monolayer MoS_2, including its internal responsivity and quantum efficiency.

Similarly, transition metal dichalcogenides (TMDs), particularly MoS_2, have garnered attention for their semiconducting applications at the nanoscale. These materials exhibit direct band gaps and high charge mobilities, making them ideal for use in nanoscale devices. One of the critical aspects of TMDs is the understanding of Schottky barrier heights

(SBHs) [103], which are crucial for efficient charge injection into these semiconductors. These theoretical models help in estimating the charge neutrality levels and canonical Schottky barrier heights, which are pivotal for the device's performance.

Further exploring the potential of MoS_2, tunneling field-effect transistors (TFETs) using monolayer MoS_2 as the channel have emerged as promising contenders for low-energy electronics. These MoS_2-TFETs [104] are compatible with silicon CMOS technology and exhibit a lower off-state leakage current and subthreshold swing compared to traditional MOSFETs.

6.3. Outlook

Future research on GO and RGO should focus on gaining a deeper understanding of the reduction mechanisms, as well as developing strategies to control both the oxidation of graphite and its subsequent reduction. After producing RGO, it can be chemically functionalized or paired with other 2D materials to tailor its properties for specialized commercial uses. For instance, the fabrication of nanocomposites, including RGO and TMDs, such as MoS_2, holds promise for supercapacitor applications.

Moreover, one of the critical areas requiring more in-depth study is the understanding of strain in graphene [105], especially when grown on transition metals via CVD or when mechanically transferred in the case of exfoliated graphene. The methods used for deposition or synthesis can have a significant impact on how graphene interacts with its substrate. Therefore, future studies should aim to establish a connection between the strength of this interaction and the thermal evolution of graphene's Raman spectra. Although some efforts have been made to correlate strain with structural features, such as surface morphology, additional research is needed to establish connections between the strain and graphene's electrical, optical, and thermal properties. Since doping has an impact on the strain in thin films, more research is also essential to explore its specific effects on the strain in graphene films.

Author Contributions: Writing—original draft preparation, G.G.P. and C.V.; writing—review and editing, G.G.P. and C.V.; supervision, C.V. All authors have read and agreed to the published version of the manuscript.

Funding: This research received no external funding.

Data Availability Statement: Not applicable.

Conflicts of Interest: The authors declare no conflict of interest.

References

1. Ding, H.; Zhang, X. Sodium Intercalation in Nitrogen-Doped Graphene-Based Anode: A First-Principles Study. *Crystals* **2023**, *13*, 1011. [CrossRef]
2. Sultana, S.M.N.; Helal, E.; Gutiérrez, G.; David, E.; Moghimian, N.; Demarquette, N.R. Effect of Few-Layer Graphene on the Properties of Mixed Polyolefin Waste Stream. *Crystals* **2023**, *13*, 358. [CrossRef]
3. Sartanavičius, A.; Žemgulytė, J.; Ragulis, P.; Ratautas, K.; Trusovas, R. Laser-Induced Graphene in Polyimide for Antenna Applications. *Crystals* **2023**, *13*, 1003. [CrossRef]
4. Zhu, C.; Lin, Z.; Zhang, D.; Shi, J.; Peng, S.; Jin, Z. Predicting the Level of Background Current Noise in Graphene Biosensor through a Non-Covalent Functionalization Process. *Crystals* **2023**, *13*, 359. [CrossRef]
5. Le, H.C.; Pham, N.T.; Vu, D.C.; Pham, D.L.; Nguyen, S.H.; Nguyen, T.T.; Nguyen, C.D. Nitrogen-Doped Graphene Quantum Dot–Tin Dioxide Nanocomposite Ultrathin Films as Efficient Electron Transport Layers for Planar Perovskite Solar Cells. *Crystals* **2023**, *13*, 961. [CrossRef]
6. Wang, Y.; Zou, B.; Rente, B.; Alford, N.; Petrov, P.K. Deposition of Nanocrystalline Multilayer Graphene Using Pulsed Laser Deposition. *Crystals* **2023**, *13*, 881. [CrossRef]
7. Hui, Y.; Zu, H.; Song, R.; Fu, H.; Luo, K.; Tian, C.; Wu, B.; Huang, G.-L.; Kou, Z.; Cheng, X.; et al. Graphene-Assembled Film-Based Reconfigurable Filtering Antenna with Enhanced Corrosion-Resistance. *Crystals* **2023**, *13*, 747. [CrossRef]
8. Jibin, K.P.; Augustine, S.; Velayudhan, P.; George, J.S.; Krishnageham Sidharthan, S.; Poulose, S.V.; Thomas, S. Unleashing the Power of Graphene-Based Nanomaterials for Chromium(VI) Ion Elimination from Water. *Crystals* **2023**, *13*, 1047. [CrossRef]
9. Bhuyan, M.S.A.; Uddin, M.N.; Islam, M.M.; Bipasha, F.A.; Hossain, S.S. Synthesis of graphene. *Int. Nano Lett.* **2016**, *6*, 65–83. [CrossRef]
10. Randviir, E.P.; Brownson, D.A.C.; Banks, C.E. A decade of graphene research: Production, applications and outlook. *Mater. Today* **2014**, *17*, 426–432. [CrossRef]

11. Huang, X.; Qi, X.; Boey, F.; Zhang, H. Graphene-based composites. *Chem. Soc. Rev.* **2012**, *41*, 666–686. [CrossRef] [PubMed]
12. Berry, V. Impermeability of graphene and its applications. *Carbon N. Y.* **2013**, *62*, 1–10. [CrossRef]
13. Novoselov, K.S.; Fal'ko, V.I.; Colombo, L.; Gellert, P.R.; Schwab, M.G.; Kim, K. A roadmap for graphene. *Nature* **2012**, *490*, 192–200. [CrossRef]
14. Rao, C.N.R.; Biswas, K.; Subrahmanyam, K.S.; Govindaraj, A. Graphene, the new nanocarbon. *J. Mater. Chem.* **2009**, *19*, 2457–2469. [CrossRef]
15. Stoller, M.D.; Park, S.; Zhu, Y.; An, J.; Ruoff, R.S. Graphene-Based Ultracapacitors. *Nano Lett.* **2008**, *8*, 3498–3502. [CrossRef] [PubMed]
16. Lui, C.H.; Liu, L.; Mak, K.F.; Flynn, G.W.; Heinz, T.F. Ultraflat graphene. *Nature* **2009**, *462*, 339–341. [CrossRef]
17. Bai, J.; Zhong, X.; Jiang, S.; Huang, Y.; Duan, X. Graphene nanomesh. *Nat. Nanotechnol.* **2010**, *5*, 190–194. [CrossRef]
18. Obraztsova, E.D.; Rybin, M.G.; Obraztsov, P.A. 7—Optical properties of graphene. In *Woodhead Publishing Series in Electronic and Optical Materials*; Skakalova, V., Kaiser, A.B.B.T.-G., Eds.; Woodhead Publishing: Sawston, UK, 2021; pp. 133–142. ISBN 978-0-08-102848-3. [CrossRef]
19. Jafari, B.; Gholizadeh, E.; Golmohammadi, S.; Ebadzadeh, M.; Soofi, H.; Aghili, S. An Innovative method for adjustable broadband THz to Mid-IR optical modulator using Graphene Gratings surface plasmon Fabry–Perot resonances with low insertion loss, high speed and modulation depth. *Opt. Commun.* **2023**, *530*, 129200. [CrossRef]
20. Rogalski, A. Graphene-based materials in the infrared and terahertz detector families: A tutorial. *Adv. Opt. Photonics* **2019**, *11*, 314–379. [CrossRef]
21. Mbayachi, V.B.; Ndayiragije, E.; Sammani, T.; Taj, S.; Mbuta, E.R.; Khan, A. ullah Graphene synthesis, characterization and its applications: A review. *Results Chem.* **2021**, *3*, 100163. [CrossRef]
22. Chen, X.; Zhang, L.; Chen, S. Large area CVD growth of graphene. *Synth. Met.* **2015**, *210*, 95–108. [CrossRef]
23. Shafi, M.; Duan, P.; Liu, W.; Zhang, W.; Zhang, C.; Hu, X.; Zha, Z.; Liu, R.; Liu, C.; Jiang, S.; et al. SERS Sensing Using Graphene-Covered Silver Nanoparticles and Metamaterials for the Detection of Thiram in Soil. *Langmuir* **2022**, *38*, 16183–16193. [CrossRef] [PubMed]
24. Cao, S.; Wang, T.; Sun, Q.; Tang, Y.; Hu, B.; Levy, U.; Yu, W. Graphene–Silver Hybrid Metamaterial for Tunable and High Absorption at Mid-Infrared Waveband. *IEEE Photonics Technol. Lett.* **2018**, *30*, 475–478. [CrossRef]
25. Shaban, N.; Prajapati, Y.K.; Mohammadkhani, R. Performance enhancement of waveguide-coupled and metamaterial surface plasmon resonance sensors based on silver-bismuth ferrite and graphene. *J. Mater. Sci. Mater. Electron.* **2023**, *34*, 309. [CrossRef]
26. Kaushik, V.; Kagdada, H.L.; Singh, D.K.; Pathak, S. Enhancement of SERS effect in Graphene-Silver hybrids. *Appl. Surf. Sci.* **2022**, *574*, 151724. [CrossRef]
27. Yu, X.; Cai, H.; Zhang, W.; Li, X.; Pan, N.; Luo, Y.; Wang, X.; Hou, J.G. Tuning Chemical Enhancement of SERS by Controlling the Chemical Reduction of Graphene Oxide Nanosheets. *ACS Nano* **2011**, *5*, 952–958. [CrossRef]
28. Zhang, J.; Yang, H.; Shen, G.; Cheng, P.; Zhang, J.; Guo, S. Reduction of graphene oxide via L-ascorbic acid. *Chem. Commun.* **2010**, *46*, 1112–1114. [CrossRef]
29. Huq, T.; Ong, H.C.; Chew, B.T.; Leong, K.Y.; Kazi, S.N. Review on aqueous graphene nanoplatelet Nanofluids: Preparation, Stability, thermophysical Properties, and applications in heat exchangers and solar thermal collectors. *Appl. Therm. Eng.* **2022**, *210*, 118342. [CrossRef]
30. Ajala, O.J.; Tijani, J.O.; Bankole, M.T.; Abdulkareem, A.S. A critical review on graphene oxide nanostructured material: Properties, Synthesis, characterization and application in water and wastewater treatment. *Environ. Nanotechnol. Monit. Manag.* **2022**, *18*, 100673. [CrossRef]
31. Razaq, A.; Bibi, F.; Zheng, X.; Papadakis, R.; Jafri, S.H.; Li, H. Review on Graphene-, Graphene Oxide-, Reduced Graphene Oxide-Based Flexible Composites: From Fabrication to Applications. *Materials* **2022**, *15*, 1012. [CrossRef]
32. Chen, D.; Feng, H.; Li, J. Graphene oxide: Preparation, functionalization, and electrochemical applications. *Chem. Rev.* **2012**, *112*, 6027–6053. [CrossRef]
33. Yang, Q.-D.; Li, J.; Cheng, Y.; Li, H.-W.; Guan, Z.; Yu, B.; Tsang, S.-W. Graphene oxide as an efficient hole-transporting material for high-performance perovskite solar cells with enhanced stability. *J. Mater. Chem. A* **2017**, *5*, 9852–9858. [CrossRef]
34. De Silva, K.K.H.; Huang, H.-H.; Joshi, R.K.; Yoshimura, M. Chemical reduction of graphene oxide using green reductants. *Carbon N. Y.* **2017**, *119*, 190–199. [CrossRef]
35. Yee, K.; Ghayesh, M.H. A review on the mechanics of graphene nanoplatelets reinforced structures. *Int. J. Eng. Sci.* **2023**, *186*, 103831. [CrossRef]
36. Moosa, A.; Ramazani, S.A.A.; Ibrahim, M. Mechanical and Electrical Properties of Graphene Nanoplates and Carbon-Nanotubes Hybrid Epoxy Nanocomposites. *Am. J. Mater. Sci.* **2016**, *6*, 157–165. [CrossRef]
37. Kumar, D.; Singh, K.; Verma, V.; Bhatti, H.S. Microwave assisted synthesis and characterization of graphene nanoplatelets. *Appl. Nanosci.* **2016**, *6*, 97–103. [CrossRef]
38. Krittayavathananon, A.; Li, X.; Batchelor-McAuley, C.; Sawangphruk, M.; Compton, R.G. Comparing the effect of different surfactants on the aggregation and electrical contact properties of graphene nanoplatelets. *Appl. Mater. Today* **2018**, *12*, 163–167. [CrossRef]

39. Moriche, R.; Jiménez-Suárez, A.; Sánchez, M.; Prolongo, S.G.; Ureña, A. High sensitive damage sensors based on the use of functionalized graphene nanoplatelets coated fabrics as reinforcement in multiscale composite materials. *Compos. Part B Eng.* **2018**, *149*, 31–37. [CrossRef]
40. Manzeli, S.; Ovchinnikov, D.; Pasquier, D.; Yazyev, O.V.; Kis, A. 2D transition metal dichalcogenides. *Nat. Rev. Mater.* **2017**, *2*, 17033. [CrossRef]
41. Han, S.A.; Bhatia, R.; Kim, S.-W. Synthesis, properties and potential applications of two-dimensional transition metal dichalcogenides. *Nano Converg.* **2015**, *2*, 17. [CrossRef]
42. Babu, G.; Masurkar, N.; Al Salem, H.; Arava, L.M.R. Transition Metal Dichalcogenide Atomic Layers for Lithium Polysulfides Electrocatalysis. *J. Am. Chem. Soc.* **2017**, *139*, 171–178. [CrossRef] [PubMed]
43. Chen, B.; Chao, D.; Liu, E.; Jaroniec, M.; Zhao, N.; Qiao, S.-Z. Transition metal dichalcogenides for alkali metal ion batteries: Engineering strategies at the atomic level. *Energy Environ. Sci.* **2020**, *13*, 1096–1131. [CrossRef]
44. Politano, G.G.; Versace, C.; Vena, C.; Castriota, M.; Ciuchi, F.; Fasanella, A.; Desiderio, G.; Cazzanelli, E. Physical investigation of electrophoretically deposited graphene oxide and reduced graphene oxide thin films. *J. Appl. Phys.* **2016**, *120*, 195307. [CrossRef]
45. Politano, G.G.; Cazzanelli, E.; Versace, C.; Vena, C.; De Santo, M.P.; Castriota, M.; Ciuchi, F.; Bartolino, R. Graphene oxide on magnetron sputtered silver thin films for SERS and metamaterial applications. *Appl. Surf. Sci.* **2018**, *427*, 927–933. [CrossRef]
46. Politano, G.G.; Cazzanelli, E.; Versace, C.; Castriota, M.; Desiderio, G.; Davoli, M.; Vena, C.; Bartolino, R. Micro-Raman investigation of Ag/graphene oxide/Au sandwich structure. *Mater. Res. Express* **2019**, *6*, 075605. [CrossRef]
47. Politano, G.G.; Nucera, A.; Castriota, M.; Desiderio, G.; Vena, C.; Versace, C. Spectroscopic and morphological study of graphene nanoplatelets thin films on Si/SiO2 substrates. *Mater. Res. Express* **2019**, *6*, 106432. [CrossRef]
48. Politano, G.G.; Castriota, M.; De Santo, M.P.; Pipita, M.M.; Desiderio, G.; Vena, C.; Versace, C. Variable angle spectroscopic ellipsometry characterization of spin-coated MoS2 films. *Vacuum* **2021**, *189*, 110232. [CrossRef]
49. Castriota, M.; Politano, G.G.; Vena, C.; De Santo, M.P.; Desiderio, G.; Davoli, M.; Cazzanelli, E.; Versace, C. Variable Angle Spectroscopic Ellipsometry investigation of CVD-grown monolayer graphene. *Appl. Surf. Sci.* **2019**, *467–468*, 213–220. [CrossRef]
50. Bruna, M.; Ott, A.K.; Ijäs, M.; Yoon, D.; Sassi, U.; Ferrari, A.C. Doping Dependence of the Raman Spectrum of Defected Graphene. *ACS Nano* **2014**, *8*, 7432–7441. [CrossRef]
51. Bonaccorso, F.; Sun, Z.; Hasan, T.; Ferrari, A.C. Graphene photonics and optoelectronics. *Nat. Photonics* **2010**, *4*, 611. [CrossRef]
52. Katsnelson, M.I. Graphene: Carbon in two dimensions. *Mater Today* **2007**, *10*, 20. [CrossRef]
53. Fang, W.; Song, Y.; Hsu, A.L.; Kong, J. A review of large-area bilayer graphene synthesis by chemical vapor deposition. *Nanoscale* **2015**, *7*, 20335. [CrossRef]
54. Yu, X.Z.; Hwang, C.G.; Jozwiak, C.M.; Köhl, A.; Schmid, A.K.; Lanzara, A. New synthesis method for the growth of epitaxial graphene. *J. Electron Spectros. Relat. Phenomena* **2011**, *184*, 100–106. [CrossRef]
55. Sinclair, R.C.; Suter, J.L.; Coveney, P. V Micromechanical exfoliation of graphene on the atomistic scale. *Phys. Chem. Chem. Phys.* **2019**, *21*, 5716–5722. [CrossRef]
56. Mishra, N.; Boeckl, J.; Motta, N.; Iacopi, F. Graphene growth on silicon carbide: A review. *Phys. Status Solidi* **2016**, *213*, 2277–2289. [CrossRef]
57. Thakur, S.; Karak, N. Alternative methods and nature-based reagents for the reduction of graphene oxide: A review. *Carbon N. Y.* **2015**, *94*, 224–242. [CrossRef]
58. Pei, S.; Cheng, H.-M. The reduction of graphene oxide. *Carbon N. Y.* **2012**, *50*, 3210–3228. [CrossRef]
59. Boukhvalov, D.W.; Katsnelson, M.I. Modeling of Graphite Oxide. *J. Am. Chem. Soc.* **2008**, *130*, 10697–10701. [CrossRef] [PubMed]
60. Nakajima, T.; Matsuo, Y. Formation process and structure of graphite oxide. *Carbon N. Y.* **1994**, *32*, 469–475. [CrossRef]
61. Smirnov, V.A.; Arbuzov, A.A.; Shul'ga, Y.M.; Baskakov, S.A.; Martynenko, V.M.; Muradyan, V.E.; Kresova, E.I. Photoreduction of graphite oxide. *High Energy Chem.* **2011**, *45*, 57–61. [CrossRef]
62. Petridis, L.V.; Kokkinos, N.C.; Mitropoulos, A.C.; Kyzas, G.Z. Chapter 8—Graphene aerogels for oil absorption. In *Advanced Low-Cost Separation Techniques in Interface Science*; Kyzas, G.Z., Mitropoulos, A.C., Eds.; Elsevier: Amsterdam, The Netherlands, 2019; Volume 30, pp. 173–197. ISBN 1573-4285. [CrossRef]
63. Konkena, B.; Vasudevan, S. Understanding Aqueous Dispersibility of Graphene Oxide and Reduced Graphene Oxide through pKa Measurements. *J. Phys. Chem. Lett.* **2012**, *3*, 867–872. [CrossRef] [PubMed]
64. Paredes, J.I.; Villar-Rodil, S.; Martínez-Alonso, A.; Tascón, J.M.D. Graphene Oxide Dispersions in Organic Solvents. *Langmuir* **2008**, *24*, 10560–10564. [CrossRef] [PubMed]
65. Yu, H.; Zhang, B.; Bulin, C.; Li, R.; Xing, R. High-efficient Synthesis of Graphene Oxide Based on Improved Hummers Method. *Sci. Rep.* **2016**, *6*, 36143. [CrossRef] [PubMed]
66. Dreyer, D.R.; Park, S.; Bielawski, C.W.; Ruoff, R.S. The chemistry of graphene oxide. *Chem. Soc. Rev.* **2010**, *39*, 228–240. [CrossRef]
67. Kwon, S.; Lee, K.E.; Lee, H.; Koh, S.J.; Ko, J.-H.; Kim, Y.-H.; Kim, S.O.; Park, J.Y. The Effect of Thickness and Chemical Reduction of Graphene Oxide on Nanoscale Friction. *J. Phys. Chem. B* **2018**, *122*, 543–547. [CrossRef]
68. Zhao, C.; Xing, L.; Xiang, J.; Cui, L.; Jiao, J.; Sai, H.; Li, Z.; Li, F. Formation of uniform reduced graphene oxide films on modified PET substrates using drop-casting method. *Particuology* **2014**, *17*, 66–73. [CrossRef]
69. Sun, H.B.; Yang, J.; Zhou, Y.Z.; Zhao, N.; Li, D. Preparation of reduced graphene oxide films by dip coating technique and their electrical conductivity. *Mater. Technol.* **2014**, *29*, 14–20. [CrossRef]

70. Nasiłowska, B.; Bogdanowicz, Z.; Hińcza, K.; Mierczyk, Z.; Góźdź, S.; Djas, M.; Kowiorski, K.; Bombalska, A.; Kowalik, A. Graphene Oxide Aerosol Deposition and its Influence on Cancer Cells. Preliminary Results. *Materials* **2020**, *13*, 4464. [CrossRef]
71. Komarov, I.A.; Danilova, E.A.; Denisenko, E.I.; Peretiyagin, V.G.; Rabchinskii, M.K. Spin-coating deposition of graphene oxide from mixed water-organic suspensions. *Fuller. Nanotub. Carbon Nanostruct.* **2022**, *30*, 146–151. [CrossRef]
72. Marcin Behunová, D.; Gallios, G.; Girman, V.; Kolev, H.; Kaňuchová, M.; Dolinská, S.; Václavíková, M. Electrophoretic Deposition of Graphene Oxide on Stainless Steel Substrate. *Nanomaterials* **2021**, *11*, 1779. [CrossRef]
73. Li, X.; Shen, Q.; Zhang, Y.; Wang, L.; Nie, C. Wear behavior of electrodeposited nickel/graphene composite coating. *Diam. Relat. Mater.* **2021**, *119*, 108589. [CrossRef]
74. Tseluikin, V.; Dzhumieva, A.; Tikhonov, D.; Yakovlev, A.; Strilets, A.; Tribis, A.; Lopukhova, M. Pulsed Electrodeposition and Properties of Nickel-Based Composite Coatings Modified with Graphene Oxide. *Coatings* **2022**, *12*, 656. [CrossRef]
75. Ray, S.C. Applications of Graphene and Graphene-Oxide based Nanomaterials. In *Applications of Graphene and Graphene-Oxide Based Nanomaterials*; Elsevier: Amsterdam, The Netherlands, 2015; pp. 1–84.
76. Tkachev, S.V.; Buslaeva, E.Y.; Naumkin, A.V.; Kotova, S.L.; Laure, I.V.; Gubin, S.P. Reduced graphene oxide. *Inorg. Mater.* **2012**, *48*, 796–802. [CrossRef]
77. Park, S.; An, J.; Potts, J.R.; Velamakanni, A.; Murali, S.; Ruoff, R.S. Hydrazine-reduction of graphite-and graphene oxide. *Carbon N. Y.* **2011**, *49*, 3019–3023. [CrossRef]
78. Fernández-Merino, M.J.; Guardia, L.; Paredes, J.I.; Villar-Rodil, S.; Solís-Fernández, P.; Martínez-Alonso, A.; Tascón, J.M.D. Vitamin C Is an Ideal Substitute for Hydrazine in the Reduction of Graphene Oxide Suspensions. *J. Phys. Chem. C* **2010**, *114*, 6426–6432. [CrossRef]
79. Jung, I.; Dikin, D.A.; Piner, R.D.; Ruoff, R.S. Tunable Electrical Conductivity of Individual Graphene Oxide Sheets Reduced at "Low" Temperatures. *Nano Lett.* **2008**, *8*, 4283–4287. [CrossRef]
80. Stankovich, S.; Dikin, D.A.; Piner, R.D.; Kohlhaas, K.A.; Kleinhammes, A.; Jia, Y.; Wu, Y.; Nguyen, S.T.; Ruoff, R.S. Synthesis of graphene-based nanosheets via chemical reduction of exfoliated graphite oxide. *Carbon N. Y.* **2007**, *45*, 1558–1565. [CrossRef]
81. Sadhukhan, S.; Ghosh, T.K.; Rana, D.; Roy, I.; Bhattacharyya, A.; Sarkar, G.; Chakraborty, M.; Chattopadhyay, D. Studies on synthesis of reduced graphene oxide (RGO) via green route and its electrical property. *Mater. Res. Bull.* **2016**, *79*, 41–51. [CrossRef]
82. Chua, C.K.; Pumera, M. Chemical reduction of graphene oxide: A synthetic chemistry viewpoint. *Chem. Soc. Rev.* **2014**, *43*, 291–312. [CrossRef]
83. Zhu, Y.; Murali, S.; Cai, W.; Li, X.; Suk, J.W.; Potts, J.R.; Ruoff, R.S. Graphene and graphene oxide: Synthesis, properties, and applications. *Adv. Mater.* **2010**, *22*, 3906–3924. [CrossRef]
84. Ferrari, A.C. Raman spectroscopy of graphene and graphite: Disorder, electron–phonon coupling, doping and nonadiabatic effects. *Solid State Commun.* **2007**, *143*, 47–57. [CrossRef]
85. Ferrari, A.C.; Meyer, J.C.; Scardaci, V.; Casiraghi, C.; Lazzeri, M.; Mauri, F.; Piscanec, S.; Jiang, D.; Novoselov, K.S.; Roth, S.; et al. Raman Spectrum of Graphene and Graphene Layers. *Phys. Rev. Lett.* **2006**, *97*, 187401. [CrossRef] [PubMed]
86. Das, A.; Pisana, S.; Chakraborty, B.; Piscanec, S.; Saha, S.K.; Waghmare, U.V.; Novoselov, K.S.; Krishnamurthy, H.R.; Geim, A.K.; Ferrari, A.C.; et al. Monitoring dopants by Raman scattering in an electrochemically top-gated graphene transistor. *Nat. Nanotechnol.* **2008**, *3*, 210–215. [CrossRef]
87. Sharma, N.; Tomar, S.; Shkir, M.; Kant Choubey, R.; Singh, A. Study of Optical and Electrical Properties of Graphene Oxide. *Mater. Today Proc.* **2021**, *36*, 730–735. [CrossRef]
88. Wu, J.-B.; Lin, M.-L.; Cong, X.; Liu, H.-N.; Tan, P.-H. Raman spectroscopy of graphene-based materials and its applications in related devices. *Chem. Soc. Rev.* **2018**, *47*, 1822–1873. [CrossRef] [PubMed]
89. De Silva, K.K.H.; Viswanath, P.; Rao, V.K.; Suzuki, S.; Yoshimura, M. New Insight into the Characterization of Graphene Oxide and Reduced Graphene Oxide Monolayer Flakes on Si-Based Substrates by Optical Microscopy and Raman Spectroscopy. *J. Phys. Chem. C* **2021**, *125*, 7791–7798. [CrossRef]
90. Malard, L.M.; Pimenta, M.A.; Dresselhaus, G.; Dresselhaus, M.S. Raman spectroscopy in graphene. *Phys. Rep.* **2009**, *473*, 51–87. [CrossRef]
91. Verde, M.; Peiteado, M.; Caballero, A.C.; Villegas, M.; Ferrari, B. Electrophoretic Deposition of Transparent ZnO Thin Films from Highly Stabilized Colloidal Suspensions. *J. Colloid Interface Sci.* **2012**, *373*, 27–33. [CrossRef]
92. Yan, Y.; Peng, S.; Jin, Z.; Zhang, D.; Shi, J. Recovering the Intrinsic Electrical Property of a Graphene Field-Effect Transistor by Interface Cleaning Technology. *ACS Appl. Electron. Mater.* **2023**, *5*, 3113–3119. [CrossRef]
93. Kim, S.; Kim, B.; Park, S.; Chang, W.S.; Kang, H.; Kim, S.; Lee, H.; Kim, S. Hysteretic behavior of all CVD h-BN/graphene/h-BN heterostructure field-effect transistors by interfacial charge trap. *Surf. Interfaces* **2023**, *36*, 102615. [CrossRef]
94. Chaudhary, M.; Xin, C.; Hu, Z.; Zhang, D.; Radtke, G.; Xu, X.; Billot, L.; Tripon-Canseliet, C.; Chen, Z. Nitrogen-Doped Carbon Quantum Dots on Graphene for Field-Effect Transistor Optoelectronic Memories. *Adv. Electron. Mater.* **2023**, *9*, 2300159. [CrossRef]
95. Khan, M.F.; Elahi, E.; Hassan, N.U.; Rehman, M.A.; Khalil, H.M.W.; Khan, M.A.; Rehman, S.; Hao, A.; Noh, H.; Khan, K.; et al. Bipolar Photoresponse of a Graphene Field-Effect Transistor Induced by Photochemical Reactions. *ACS Appl. Electron. Mater.* **2023**. [CrossRef]
96. Vignesh, G.; Devendran, P.; Nallamuthu, N.; Sudhahar, S.; Kumar, P.S.; Kumar, M.K. Effects of nitrogen, sulphur, and temperature treatments on the spectral, structural, and electrochemical characteristics of graphene oxide for energy storage applications. *Carbon Trends* **2023**, *11*, 100262. [CrossRef]

97. Zhang, D.; Du, J.; Zhang, W.; Tong, B.; Sun, Y.; Zhao, T.-Y.; Ma, L.-P.; Sun, D.-M.; Cheng, H.-M.; Ren, W. Carrier Transport Regulation of Pixel Graphene Transparent Electrodes for Active-Matrix Organic Light-Emitting Diode Display. *Small* **2023**, 2302920. [CrossRef] [PubMed]
98. Fan, X.; Smith, A.D.; Forsberg, F.; Wagner, S.; Schröder, S.; Akbari, S.S.A.; Fischer, A.C.; Villanueva, L.G.; Östling, M.; Lemme, M.C.; et al. Manufacture and characterization of graphene membranes with suspended silicon proof masses for MEMS and NEMS applications. *Microsyst. Nanoeng.* **2020**, *6*, 17. [CrossRef] [PubMed]
99. Zhang, M.; Han, N.; Zhang, J.; Wang, J.; Chen, X.; Zhao, J.; Gan, X. Emergent second-harmonic generation in van der Waals heterostructure of bilayer MoS2 and monolayer graphene. *Sci. Adv.* **2023**, *9*, eadf4571. [CrossRef]
100. Thakkar, S.; De Luca, L.; Gaspa, S.; Mariani, A.; Garroni, S.; Iacomini, A.; Stagi, L.; Innocenzi, P.; Malfatti, L. Comparative Evaluation of Graphene Nanostructures in GERS Platforms for Pesticide Detection. *ACS Omega* **2022**, *7*, 5670–5678. [CrossRef]
101. Szczęśniak, D.; Kais, S. Gap states and valley-spin filtering in transition metal dichalcogenide monolayers. *Phys. Rev. B* **2020**, *101*, 115423. [CrossRef]
102. Pak, J.; Lee, I.; Cho, K.; Kim, J.-K.; Jeong, H.; Hwang, W.-T.; Ahn, G.H.; Kang, K.; Yu, W.J.; Javey, A.; et al. Intrinsic Optoelectronic Characteristics of MoS2 Phototransistors via a Fully Transparent van der Waals Heterostructure. *ACS Nano* **2019**, *13*, 9638–9646. [CrossRef]
103. Szczęśniak, D.; Hoehn, R.D.; Kais, S. Canonical Schottky barrier heights of transition metal dichalcogenide monolayers in contact with a metal. *Phys. Rev. B* **2018**, *97*, 195315. [CrossRef]
104. Lan, Y.-W.; Torres, C.M., Jr.; Tsai, S.-H.; Zhu, X.; Shi, Y.; Li, M.-Y.; Li, L.-J.; Yeh, W.-K.; Wang, K.L. Atomic-Monolayer MoS2 Band-to-Band Tunneling Field-Effect Transistor. *Small* **2016**, *12*, 5676–5683. [CrossRef] [PubMed]
105. Ferralis, N. Probing Mechanical Properties of Graphene with Raman Spectroscopy. *J. Mater. Sci.* **2010**, *45*, 5135–5149. [CrossRef]

Disclaimer/Publisher's Note: The statements, opinions and data contained in all publications are solely those of the individual author(s) and contributor(s) and not of MDPI and/or the editor(s). MDPI and/or the editor(s) disclaim responsibility for any injury to people or property resulting from any ideas, methods, instructions or products referred to in the content.

crystals

MDPI

Article

Unleashing the Power of Graphene-Based Nanomaterials for Chromium(VI) Ion Elimination from Water

Keloth Paduvilan Jibin [1,*,†], Silpa Augustine [2,†], Prajitha Velayudhan [1], Jesiya Susan George [1], Sisanth Krishnageham Sidharthan [3], Sylas Variyattel Paulose [4,*] and Sabu Thomas [1,3,*]

1 School of Chemical Sciences, Mahatma Gandhi University, Kottayam 686560, India; prajitha@mgu.ac.in (P.V.); jesiya@mgu.ac.in (J.S.G.)
2 Department of Chemistry, Baselius College, Kottayam 686560, India; srsilpanov@gmail.com
3 International and Inter-University Centre for Nano Science and Nanotechnology, Mahatma Gandhi University, Kottayam 686560, India; sisanth@mgu.ac.in
4 School of Environmental Sciences, Mahatma Gandhi University, Kottayam 686560, India
* Correspondence: jibin@mgu.ac.in (K.P.J.); sylas@mgu.ac.in (S.V.P.); sabuthomas@mgu.ac.in (S.T.)
† These authors contributed equally to this manuscript.

Abstract: Graphene-oxide-based nanomaterials have gained attention in recent years in the field of heavy metal removal. In this work, graphene oxide (GO) and graphene-oxide-coated silica nanoparticles (GO/SiO_2) were synthesized for the efficient removal of Cr(VI) ions from water. Synthesized nanosorbents were characterized by FTIR, Raman spectroscopy, and Transmission Electron Microscopy (TEM). The effects of the pH and the concentration of Cr(VI) ions in adsorption, using GO and GO/SiO_2, was studied using the batch process. The results of the study indicated that the maximum removal percentage was shown at pH 3 for both adsorbents. Comparatively, GO/SiO_2 showed a higher removal percentage (92.28%) than GO (86.15%) for Cr(VI) at a concentration of 50 ppm. The results validate that the removal of Cr(VI) ions is highly concentration-dependent and pH-dependent. This study shows that GO and GO/SiO_2 are efficient adsorbents and that GO/SiO_2 has great potential over GO for the removal of Cr(VI) ions from water.

Keywords: heavy metals; graphene oxide (GO); graphene-oxide-coated silica nanoparticles (GO/SiO_2); adsorption; ICP-MS

1. Introduction

The availability of clean drinking water is critical for maintaining a healthy life. Freshwater resources have been steadily declining in recent decades due to increased human consumption, pollution, and climate change. Among the toxic pollutants found in water bodies, heavy metals deserve special attention. Heavy metals are generally referred to as those metals which possess a specific density of more than 5 g/cm^3 and adversely affect the environment and living organisms [1]. They are very toxic even at low doses, very persistent in the environment and living tissues, and easily migrate upstream of the food chain. The toxic effects of these metals are harmful to the human body and its proper functioning. Moreover, monitoring and removing them are costly procedures. The most commonly found heavy metals in wastewater include arsenic, cadmium, chromium, copper, lead, nickel, mercury, and zinc, all of which badly affect human health and the environment [2]. Heavy metals enter into the surrounding environment by natural means and as a result of human activities. Various sources of heavy metals include soil erosion, natural weathering of the earth's crust, mining, industrial discharge, urban runoff, sewage discharge, pest- or disease-control agents applied to crops, and many others [3].

Chromium (Cr) is the seventh most abundant element on earth [4]. It occurs in several oxidation states in the environment, ranging from Cr^{2+} to Cr^{6+}.The most commonly occurring forms of Cr are trivalent- Cr^{3+} and hexavalent- Cr^{6+}, and both states are toxic to animals, humans, and plants [5]. Hexavalent chromium is thought to have 100-fold extra toxicity than trivalent chromium for each acute and persistent exposure, due to its excessive water solubility and mobility and easy reduction. The respiratory tract is the important goal organ for hexavalent chromium after inhalation exposure in humans. Chronic human exposure to high concentrations of hexavalent chromium through inhalation or oral exposure may produce effects on the liver, kidney, gastrointestinal, and immune systems, and possibly the blood. Dermal exposure to hexavalent chromium can result in contact dermatitis, sensitivity, and ulceration of the skin [6].

Heavy metal pollution has become a serious problem in both developed and developing countries [7–10]. In addition to the detection of heavy metal ions, their elimination from water is quite a crucial concern. Many strategies can be employed for the removal process, and common methods for removing heavy metals from water sources include precipitation, flocculation, membrane separation, ion exchange, and evaporation [11–13]. Among the many methods, adsorption is often the most effective technique of choice, due to its low cost, simple design, strong operability, and especially its high elimination performance from dilute solutions [14,15]. Some widely used adsorbents for the adsorption of metal ions include activated carbon, clay minerals, biomaterials, industrial solid wastes, and zeolites [16]. The development of nanotechnology has provided various nanomaterials as efficient adsorbents for removing metal ions from water [17–20]. Heavy metal removal using graphene and graphene-based nanomaterials has grown significantly during these years. Development of various functionalized nanomaterials, such as graphene oxide (GO), enhances the removal efficiency toward metal ion uptake. The presence of oxygen-containing functional groups (hydroxyl, carboxyl, epoxy, etc.) for metal ion complexation, large specific surface area, fast adsorption speed, affinity for a large number of metal ions, and low preparation cost make GO a promising adsorbent for heavy metal removal [21,22]. Rajesh et al. reported a novel and straightforward approach for effectively removing toxic chromium by using the intriguing interaction between exfoliated graphene oxide (EGO), trioctylamine (TOA), and Cr(VI) [23]. Recently, numerous research works have studied the development of graphene-oxide-based nanomaterials and their environmental applications [24]. For instance, Deng et al. synthesized magnetic graphene oxide and used it as an adsorbent for the removal of Cd^{2+} from wastewater [25]. Likewise, a synthesized magnetic graphene oxide (mGO) grafted with amine groups (mGO-PAMAM) was found to be effective in the removal of Cd^{2+}, Pb^{2+}, and Cu^{2+} from wastewater and to have excellent regeneration capability for future use [26]. Prepared magnetic chitosan and graphene nanocomposites functionalized with EDTA (EDTA-MCS/GO) were used for the removal of Cr^{3+} from water [27]. In this study, we synthesized graphene oxide (GO),amine-modified silica using APTES and graphene-oxide-coated silica nanoparticles (GO/SiO_2), for the removal of Cr(VI) ions from water. Generally, the interaction between metal ions and the surface oxygen-containing groups plays an important role in the adsorption capacity of heavy metals on CNTs and graphene-based nanomaterials [28]. Since silica nanoparticles and their derivatives have been found to possess excellent qualities, such as a definite pore size, high surface area, strong selectivity, and adsorption capability, a lot of studies have been conducted employing them as adsorbents [29]. Further modification of nano silica and synthesized GO is possible and thereby overcomes the self-aggregation and also increases the removal efficiency. When APTES-modified silica is coated with GO, the adsorption rate is found to be increased due to its enhanced properties.

In the present work, we successfully prepared an amine-modified silica-GO self-assembled novel high-efficiency adsorbent for the successful removal of Cr(VI) from the solution and compared the result with GO. The combination of amine-modified silica and GO creates a synergistic effect that improves Cr(VI) adsorption. Graphene oxide possesses a large surface area and abundant oxygen-containing functional groups, such as hydroxyl and carboxyl groups, which can form coordination complexes with metal ions. The amine-modified silica, with its amino groups, can interact with the functional groups on GO, creating a favorable environment for Cr(VI) adsorption. The synergistic effect between the amine-modified silica and GO enhances the overall adsorption capacity for Cr(VI). The properties of the prepared adsorbent were studied using various spectroscopic and microscopic methods. We achieved a high-percentage removal of Cr(VI) in the silica-GO hybrid system.

2. Materials and Methods

2.1. Materials

Potassium dichromate ($K_2Cr_2O_7$), graphite (20-micron, synthetic CAT#282863), potassium permanganate (KMnO), Conc.H_2SO_4, sodium nitrate ($NaNO_3$), hydrochloric acid (HCl), sodium carbonate (Na_2CO_3), Nano silica (10 nm diameter, spec. surface area 175–225 m^2/g (BET), 99.8% trace metals basis. CAT#718483), 3-Aminopropyltriethoxysilane (APTES), HCl, ammonia, ethanol, and 30% H_2O_2 were purchased from Sigma Aldrich (St. Louis, MI, USA). All the chemicals were used as received, unless otherwise specified.

2.2. Methods

2.2.1. Synthesis of Graphene Oxide

Graphene oxide was synthesized by modified Hummer's method [30]. Then, 5 g of graphite, 115 mL of sulfuric acid, and 5 g of $NaNO_3$ were taken in a beaker and stirred for 30 min in an ice bath. After that, 15 g of $KMnO_4$ was added very slowly and stirred for 40 min. We then slowly raised the temperature to 35 °C for 12 h, and to this, 230 mL of distilled water was added slowly. The temperature was further raised to 90 °C for 1 h, and to this, 400 mL of distilled water and 50 mL of 30% H_2O_2 were added. Then, we removed the supernatant and washed it with 5% HCl several times to remove sulphate salt. The solution was then made alkaline by treating it with 5% Na_2CO_3. Afterward, it was centrifuged and dried it in an oven.

2.2.2. Synthesis of Amine Modified Silica

The APTES and silica nanoparticles were used to create amine modified silica nanoparticles. This was accomplished by dissolving 1 mL of APTES in 200 mL of ethanol in a beaker. This solution was then mixed with 10 g of SiO_2. The resulting mixture was sonicated and stirred vigorously for one hour. The solution was then given 150 mL of water, followed by 30 min of stirring and 15 min of sonication. The mixture was centrifuged for 20 min at 14,000 rpm, the solid residue recovered and repeatedly rinsed with ethanol. The amine modified silica nanoparticles collected by gently heating the gel-like substance for three hours at 50 °C and 0.2 mbar of reduced pressure.

2.2.3. Synthesis of Self-Assembled Graphene-Oxide-Coated Silica Nanoparticle

Synthesized graphene oxide was dispersed in distilled water. Modification of amine on nano silica was accomplished by using 3-aminopropyl triethoxysilane in ethanol/water mixture (9:1). Nano silica purchase from sigma Aldrich is modified by using 3-Aminopropyl triethoxysilane (APTES) in ethanol medium (Figure 1) and the amine modified silica was also dispersed in distilled water and prepared the self-assembled structures of graphene oxide coated silica nanoparticles by mixing along with stirring and the core shell precipitates out and washed and dried. Then,10 g of amine-modified silica was dispersed in 500 mL distilled water via sonication and stirring. In another beaker, 0.1 g of graphene oxide was dispersed in 500 mL of distilled water. The two 500 mL solutions were mixed thoroughly and stirred for another 6 h. Centrifuge the precipitate and dried in vacuum oven at 80 °C.

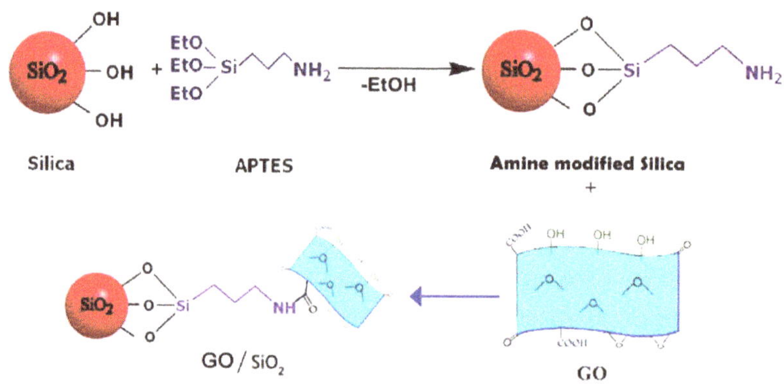

Figure 1. Synthesis of amine modified silica and GO/SiO_2.

2.2.4. Preparation of Stock Solution

Stock solution of chromium was prepared by dissolving 100 mg of potassium dichromate ($K_2Cr_2O_7$) in a 100 mL standard flask. From this stock solution of 1000 ppm, different concentrations of standard solutions (1, 10, 30, 50, 100, and 200) were prepared.

2.2.5. Optimization of pH

A series of standard solutions, 1 ppm, 10 ppm, 30 ppm, 50 ppm, 100 ppm, and 200 ppm, were prepared in a 100 mL standard flask. The pH was adjusted to 3, 5, 7, and 9 for a 30 ppm solution, and 20 mg of GO was added to each, sonicated for 20 min, and allowed to settle down for 24 h. The solutions were decanted using Whatman No. 41 filter paper, and the samples to be analyzed were filtered using a syringe filter. Then ICP-MS analysis was performed, and the results were used for determining the pH at which the maximum adsorption takes place. The same procedure was carried out using silica-coated GO for a 30 ppm standard solution, and the optimum pH was determined.

2.2.6. Batch Adsorption Test on GO and GO-Coated Silica Nanoparticles

To investigate the adsorption of heavy metal ions (Cr (VI)) on GO and GO-coated silica nanocomposites, a batch adsorption test was performed. First, 20 mg of GO was added to 20 mL of different concentrations of standard solutions (1 ppm, 10 ppm, 30 ppm, 50 ppm, 100 ppm, and 200 ppm). Then, 20 mg of GO-coated silica nanoparticles was added to the next set of 20 mL standard solutions. The pH was optimized, then the sample was sonicated for 20 min, allowed to cool at room temperature, and it was left at room temperature for 24 h for equilibration. It was then decanted using Whatman No.1 filter paper, and the sample to be analyzed was filtered using a syringe filter. Then ICP-MS analysis was carried out for the filtrate that was obtained after sonication, filtration, and syringe filtration. The quantity of adsorbed heavy metal ions was calculated as the percentage of the difference between concentrations of metal ions before and after adsorption as follows [31,32]:

$$\textit{Percentage Adsorption}\ (\textit{Removal}) = \frac{C_i - C_e}{C_e} \times 100$$

where C_i is the initial heavy metal concentration (ppm), and C_e is the heavy metal concentration at final equilibrium (ppm).

2.3. Characterizations

Characterization of graphite and graphene oxide was performed using FTIR spectrometer (FTIR spectrometer Perkin Elmer Spectrum 2). The Raman spectra were obtained using WITEC Alpha300 RA—Confocal Raman Microscope with AFM. Morphological studies of GO and GO-coated silica nanoparticle were performed with a transmission electron microscope (TEM—JEOL 2100, high resolution, Tokyo, Japan). Amounts of heavy metal removal from standard solutions were determined by using ICP-MS (iCAP Q-model-Thermo).

3. Results and Discussion

3.1. Characterization of GO and GO/SiO_2

FTIR measurement was employed to investigate the bonding interactions in graphite before and after the oxidation process. Figure 2 shows the FTIR spectra of graphite, graphene oxide, and GO/SiO_2.

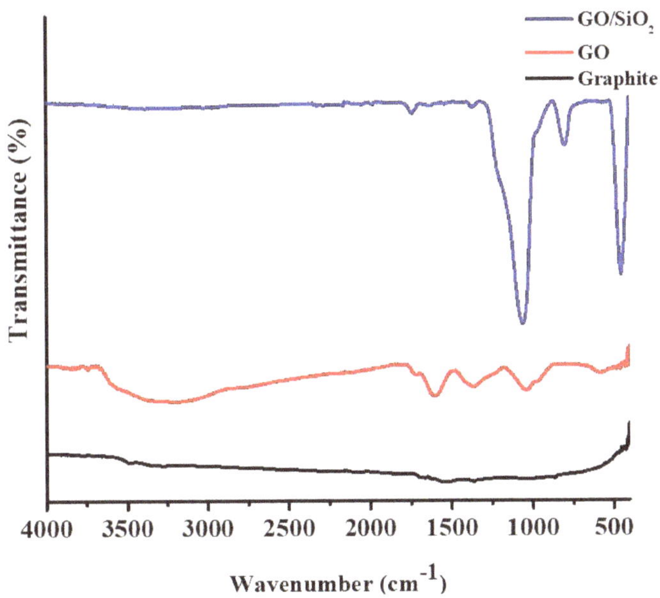

Figure 2. FTIR spectra of graphite, GO, and GO/SiO_2.

Evidence of graphite oxidation could be obtained from the analysis of spectral data. The different bands present in the spectra indicate different functional groups present on GO. Figure 2 shows that the synthesized GO has a peak between 1000 and 1100 cm^{-1} that is attributed to the epoxy (-O-) group [33]. The peak between 1700 cm^{-1} and 1720 cm^{-1} is associated with the stretching of C=O bonds of the carboxyl group (-COOH) [34] or carbonyl group [26,35]. The peak located between 3500 and 3400 cm^{-1} is ascribed to the -OH stretching mode hydroxyl group [26]. The FTIR spectrum shows oxygen-containing functional groups such as carbonyl (C=O), carboxyl—(COOH), epoxy (C-O-C), and hydroxyl (O-H) groups present on GO. The FTIR spectra of graphite do not show such properties, thereby confirming the successful incorporation of oxygen-containing groups into the graphite. From the FTIR spectra of GO-SiO_2, we can confirm the successful incorporation of SiO_2 into the GO sheets. The characteristic peaks at 1199 cm^{-1}, 1060 cm^{-1}, 970 cm^{-1}, and 800 cm^{-1} are attributed to the Si-O-C asymmetric stretching vibration Si-O-Si asymmetric stretching vibration, Si-OH stretching vibration, and Si-O-Si symmetrical stretching vibration, respectively. Moreover, the peaks were due to the Si-O-Si groups having a higher intensity than the Si-O-C and Si-OH groups. Furthermore, the peak at 1721 cm^{-1} on GO (C=O) completely disappeared, suggesting the conversion of carbonyl groups into Si-O-C bonds.

The FTIR analysis shown in Figure 3 proves the existence of modified silica nanoparticles. The bands at 810 and 475 cm^{-1} found in the FTIR spectrum of silica nanoparticles illustrated in Figure 3 can be attributed, respectively, to the stretching and bending of Si-O-Si bonds. Furthermore, siloxane vibrations from (SiO)n groups are implicated according to the absorption bands at 1100 cm^{-1}. The O-H stretching band of the surface silanol groups and the remaining adsorbed water molecules are responsible for the peaks at 3450 and 1630 cm^{-1}, respectively, which developed after the APTES modification of the silica nanoparticles (Figure 3). The bands around 2800 cm^{-1} can be connected to the asymmetrical and symmetrical stretching vibrations of CH_2 groups after silica nanoparticles were modified with APTES (Figure 3). The bending vibration of the N-H group causes the appearance of a new band at 1560 cm^{-1} in response to the addition of an amine group. Furthermore, the stretching vibration of the N-H group can be linked to the peak around 3500 cm^{-1}. These combined results show that amine-functionalized silica nanoparticles were successfully prepared.

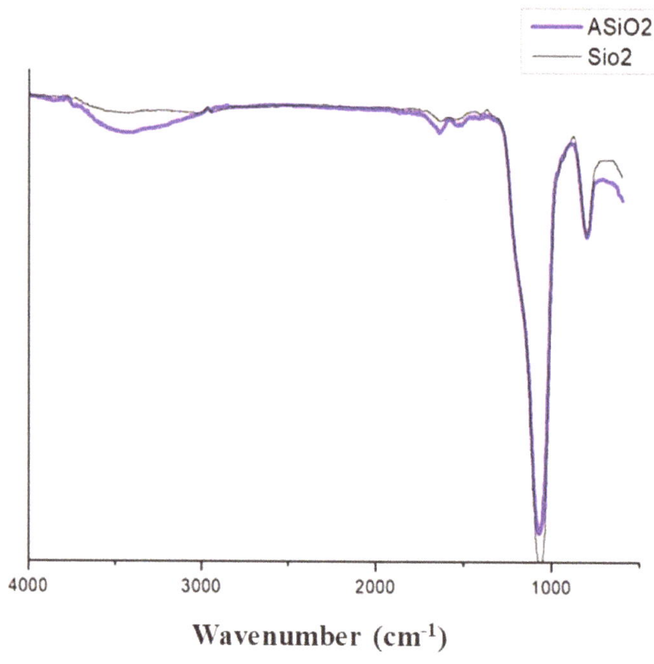

Figure 3. FTIR spectra of SiO_2 and amine-modified SiO_2.

Raman spectroscopy is a standard non-destructive technique employed for the characterization of carbon-based materials [36]. The oxidation of graphite to GO can be further confirmed by Raman spectroscopy. Characteristic peaks of graphitic carbon-based materials are the *G* and *D* bands and their overtones [37]. As shown in Figure 4, characteristic bands for graphite and GO appear at around 1300–1350 cm^{-1} (*D band*) and between 1550 and 1600 cm^{-1} (*G band*) [26,38]. Their intensity ratio (I_D/I_G) indicates the degree of the disorder and defects [38]. A universal observation is that the higher the disorder of graphite, the broader the *G* and *D bands*, which have a higher relative intensity compared to the G band [39]. As shown in Figure 3, it is observed that there is a small shift for GO, and, also, the bands are broader than they are in graphite. The *G* peak and *D* peak of GO are shifted to higher frequencies with respect to that of graphite. The increase in intensity and shifting of the *D* peak to a higher wavelength corresponds to the presence of defects and the distortion of the sp2 crystal structure of graphite [37]. The 2D band appears at 2660 cm^{-1} [40]. The intensity of the 2D band is smaller after oxidation because of the breaking of the stacking order and incorporation of functional groups due to the oxidation reaction [41]. Thus, the intense D band in GO confirms the oxygen functionalization of graphite. The ratio of I_D and I_G portrays the defect window regime in GO and GO/SiO_2. In general, the oxidation of graphite is accomplished by the increase of I_D/I_G. Hence, the higher the value of this I_D/I_G ratio, the more there will be structural defects or disorders in the system. Table 1 shows the I_D/I_G values of graphite, GO, and GO/SiO_2.

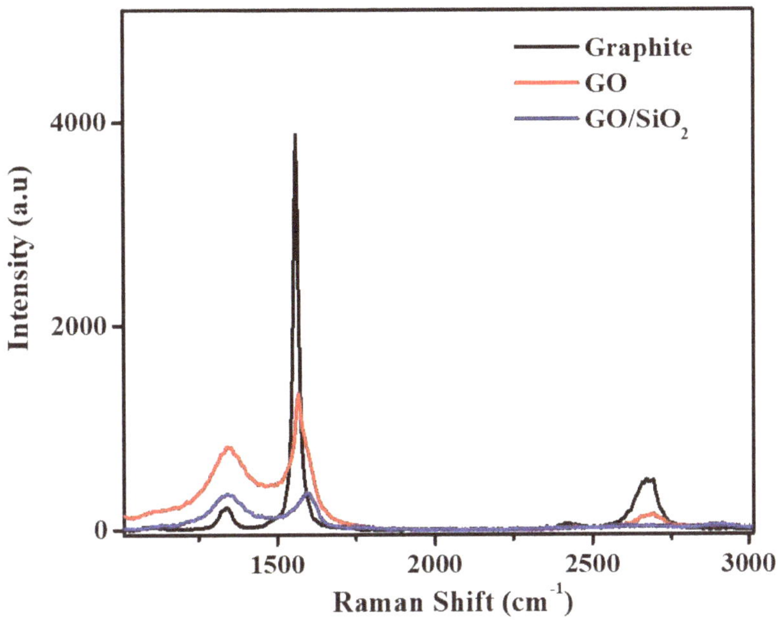

Figure 4. Raman spectra of GO and graphite.

Table 1. I_D/I_G values of graphite, GO, and GO/SiO_2.

Sample Code	I_D/I_G
Graphite	0.05
GO	0.60
GO/SiO_2	0.94

The I_D/I_G ratios of GO and GO/SiO_2 were 0.05 and 0.60, respectively. Thus, the increased value of the I_D/I_G ratio of GO/SiO_2 indicates an increased degree of disorder and improved lamellar spacing, which would assist in the better dispersion of GO-SiO_2 nanoparticles, and this result is in line with the TEM images.

Transmission electron microscopy (TEM) is considered as an effective method for the morphological studies of GO. With increases in the oxidation level, GO becomes highly transparent, since it possesses a high amount of the oxygenated functional group, which makes them suitable for exfoliation into monolayers or just a few layers of GO after ultrasonication. As shown in Figure 5, the TEM images of GO sheets are coupled with numerable wrinkles and ripples on them. The transparency level of the image indicates the presence of more layers and functionalization with oxygen-containing groups [42]. Moreover, the GO-coated silica shows a rough external surface, with visible irregular cavities (Figure 5b). This must be due to the nanoparticles embedded in the silica backbone. The TEM images containing the inner core are outer shell shows that the GO was successfully coated on the silica particles.

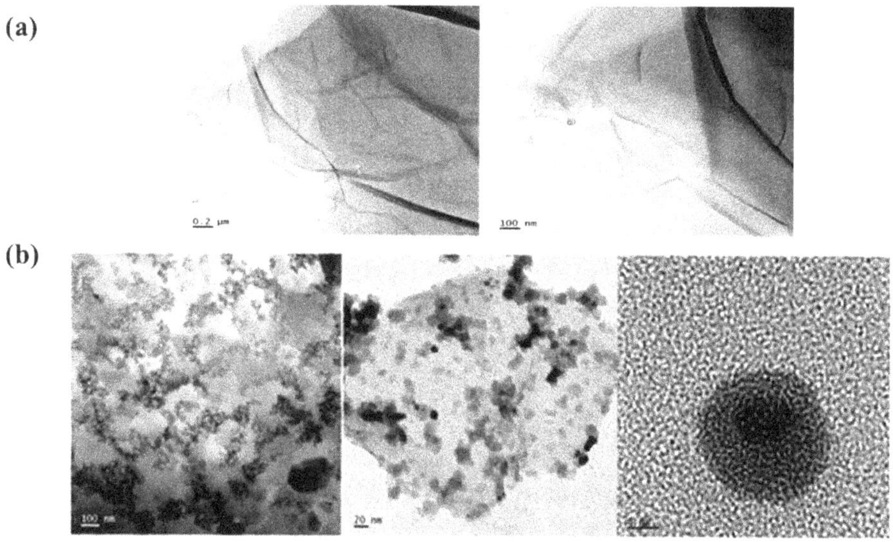

Figure 5. (**a**) Transmission electron microscopy (TEM) images of synthesized graphene oxide (GO) at different magnifications. (**b**) TEM images of graphene-oxide-coated silica in different magnifications (100 nm, 20 nm and 5 nm).

3.2. Effect of pH on Adsorption of Cr(VI) Ions

The pH is the key factor that affects the removal efficiency of metal ions. It affects the surface charge of the adsorbent, the degree of ionization, and the speciation of the adsorbate [43]. To study the effect of the pH on adsorption, experiments were conducted under the same experimental conditions, at a concentration of 30 ppm and a pH in the range of 3.0–9.0. The experimental results are shown in Table 2.

Table 2. Percentage of adsorption of Cr(VI) ions using GO and GO-coated-silica nanoparticles at 30 ppm.

pH of the Sample	Initial Concentration (C_i) (ppm)	Final Concentration (C_e) (ppm)		Adsorbance Percentage (%)	
		GO	Si/GO	GO	Si/GO
3	30	8.08331	5.77240	73.06 ± 0.6	80.76 ± 0.125
5	30	10.09871	10.97567	66.34 ± 0.3	63.41 ± 0.312
7	30	10.00400	10.68623	66.65 ± 0.025	64.38 ± 0.421
9	30	10.11465	11.00769	66.28 ± 0.532	60.97 ± 0.25

From Table 2, it is clear that the optimum pH for the efficient removal of chromium(VI) ions is pH 3. Basic medium, i.e., pH 9, leads to the least removal of Cr(VI) for both of the adsorbents (Figure 6). As reported by Zhao et al. [44] and Shirkhanloo et al. [45], at low pH values, the surface charge of GO is positive due to the protonation reaction. At lower pH values, the predominant Cr(VI) species mainly exists in the monovalent $HCrO_4^-$ form, which is then gradually changed to the divalent CrO_4^{2-} and $Cr_2O_7^{2-}$ as the pH increases [46,47]. Hence, at a lower pH, the surface of GO and GO-SiO_2 becomes protonated. Then, the positively charged surface of the adsorbents strongly attracts oppositely charged Cr(VI) species through electrostatic interaction, and the adsorption rate increases [48]. Moreover, the adsorption free energy of $HCrO_4^-$ ion is lower than that of CrO_4^{2-}, and, hence, $HCrO_4^-$ is more favorably adsorbed than CrO_4^{2-} at the same concentration [49,50]. As the pH increases, the availability of the OH^- ions' concentration increases, thus causing them to compete with the Cr(VI) ions, leading to the reduction in percentage of adsorption [32]. Thus, this study reveals that the adsorption quantities of Cr(VI) ions at a lower pH are larger than those of at a higher pH.

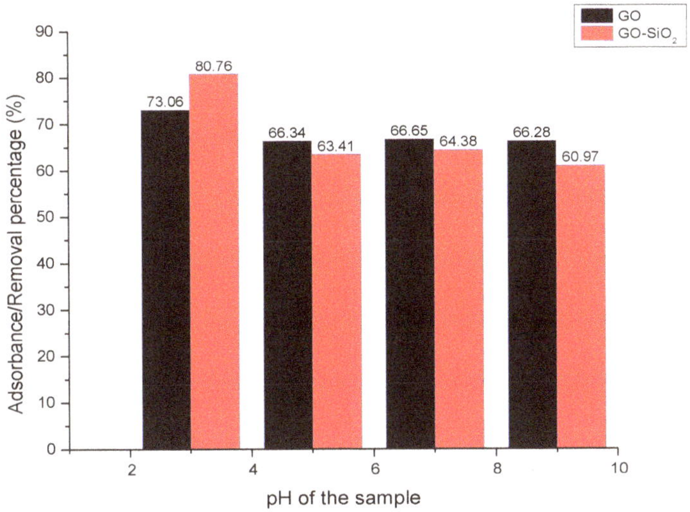

Figure 6. Effect of pH on the adsorption of Cr(VI) ion.

3.3. *Evaluation of the Adsorption Efficiency of GO and Si/GO at Different Concentrations of Solutions*

Adsorption experiments were conducted with different concentrations of heavy metal solutions (C_i, 1–200 ppm) at pH 3. Table 3 shows the effect of initial concentration of the solution under the optimum pH.

Table 3. Percentage of adsorption of Cr(VI) ions from different concentrations of samples, using GO and GO-coated silica nanoparticles.

Initial Concentration of the Sample (C_i) (ppm)	Concentration after Adsorption (C_e) (ppm)		Percentage of Adsorption (%)	
	GO	Si/GO	GO	Si/GO
1	0.61504	0.29760	38.49 ± 0.25	70.24 ± 0.65
10	5.94674	2.46630	40.53 ± 0126	75.34 ± 35
30	8.08331	5.77240	73.06 ± 0.6	80.76 ± 0.69
50	6.92333	3.85825	86.15 ± 0.8	92.28 ± 0.45
100	65.34009	52.31453	34.66 ± 0.58	47.69 ± 0.52
200	132.2878	118.8226	33.86 ± 0.87	40.59 ± 0.85

At a concentration of 50 ppm, it was noticed that the percentages of adsorption of Cr(VI) by GO and GO-SiO_2 were at the maximum, i.e., at optimum concentration; the removal percentage was 86.15% and 92.28% for GO and GO-SiO_2, respectively. As the initial concentration increases, the adsorption percentage of Cr(VI) increases and then decreases for both the adsorbents. At lower concentrations, the number of Cr(VI) ions present in the solution is comparatively lower compared to the number of available sites on the GO and GO-SiO_2 nanoparticles [51]. The highest heavy metal concentration (200 ppm) leads to the lowest removal percentage [52]. Imtithal et al. [31] reported that, at a lower concentration of metal ions, vacant sites are available for the adsorption of metal ions, and, hence, there is a higher adsorption rate. At higher concentrations, the capacity of the adsorbent becomes exhausted due to the non-availability of the surface sites; i.e., as the concentration increases, the surface sites of the adsorbents become saturated, leading to less or no more adsorption (Figure 7). Moreover, it is clear that the adsorption efficiency of GO-SiO_2 is higher than GO at all concentrations. Although the functional groups, such as epoxy (C-O-C), hydroxyl (-OH), carboxyl (-COOH), and carbonyl (C=O), on the surface of GO promote the adsorption of metal ions, the aggregation of its layers due to strong interplanar interactions leads to a reduced metal ion adsorption capacity [53]. GO and amine-modified silica are good adsorbents, and their combination leads to the formation of an excellent adsorbent, i.e., GO-SiO_2. The increases in the specific surface area and the number of binding sites of the GO-SiO_2 composite contribute to its excellent removal efficiency [54]. Both amine-modified silica and graphene oxide (GO) are good adsorbents for removing heavy metals from wastewater. Due to the abundance of functional groups such as -OH, -COOH, and -C=O, GO has a wide surface area and high adsorption capacity. These groups have high interactions with heavy metal ions through coordination and electrostatic interactions, making GO an efficient adsorbent. On the other hand, amino functional groups in amine-modified silica show potent metal-ion-chelating capabilities. It is a great adsorbent for metal ions because the amino groups can form complexes with metal ions through coordination interactions. The GO-SiO_2 composite that results from the combination of these two substances has improved the adsorption abilities for the removal of Cr(VI) from wastewater. The amine-modified silica provides chelating sites for Cr(VI) ions, whereas GO provides a high surface area for the adsorption of metal ions. A better adsorption effectiveness can also result from the presence of amine functional groups on the silica surface, which can encourage the dispersion of GO and prevent its aggregation. The interaction between protonated amine and hydro chromate anion exhibits a strong affinity for graphene oxide, primarily through cation-pi and electrostatic interactions [23,55,56]. Chromium(VI) is easily adsorbed onto the surface of GO due to hydrogen bonding and the electrostatic interaction of CrO_4^{2-} with the oxygen functionalities in GO. Figure 8 illustrates the interaction between the amine and silica, graphene oxide with amine-modified silica, and Cr(VI).

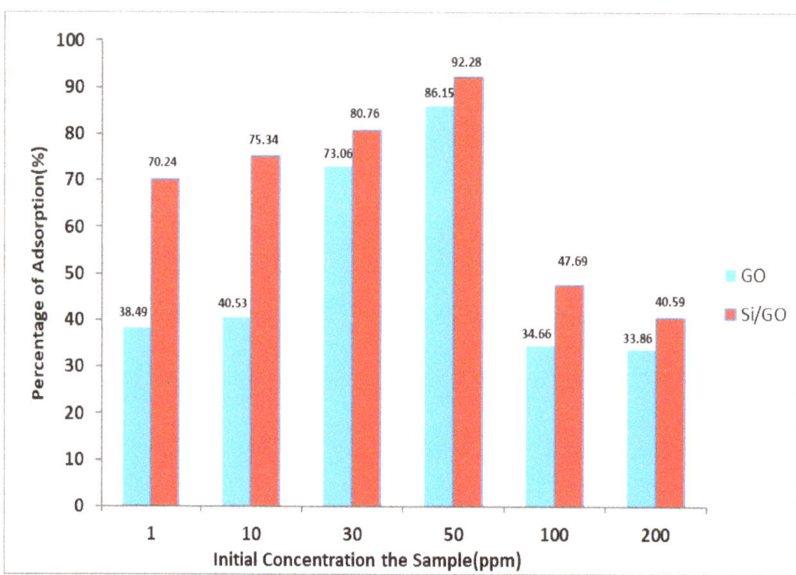

Figure 7. Adsorption of Cr(VI) using GO and Si/GO at different concentrations.

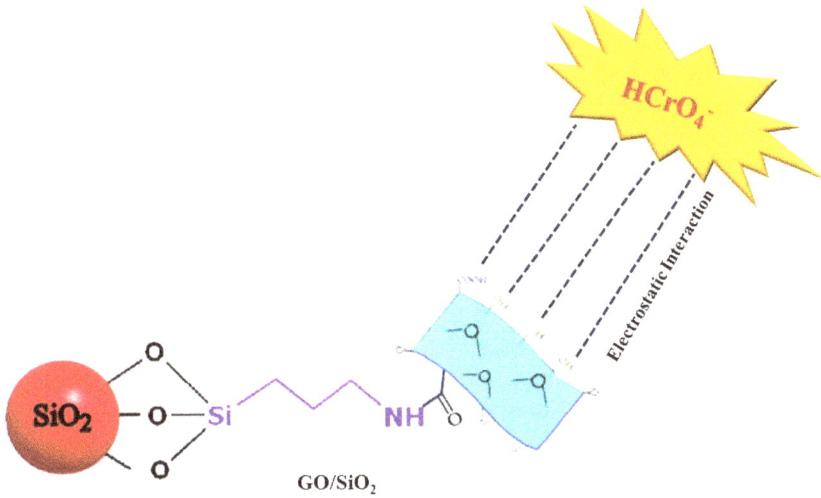

Figure 8. Schematic representation of the Cr(VI) adsorption mechanism.

3.4. Comparison of Removal Efficiency of Cr(VI) Ions Using Different Adsorbents

In order to compare the efficiency of synthesized nanoparticles for the removal of Cr(VI) ions from water with the ones reported in the literature, it is appropriate to compare the removal percentage or adsorbance capacity of graphene oxide (GO) and graphene-oxide-coated silica nanoparticles (GO/SiO_2). In this paper, a comparison of the removal percentage of different adsorbents with synthesized adsorbents is presented Table 4.

Table 4. Summary of removal percentage of Cr(VI) using various adsorbents.

No.	Adsorbent	Optimum pH	Removal % (a)/Adsorption Capacity (mg/g) (b)	Reference
1	Graphene oxide/silica nanosheets with multistage pores	2	90 (a)	[57]
2	Surface-modified Jacobsite ($MnFe_2O_4$) nanoparticles	2	31.55 (b)	[58]
3	GO prepared from Graphite	4	92.8 (a)	[32]
4	Maghemite nanoparticles	2.5	97.3 (a)	[59]
5	Magnetite nanoparticle	2.5	99.8 (a)	[49]
6	Nanostructured graphite oxide	3	63 (a)	[31]
7	Silica/graphite oxide composite	3	60.24 (b)	[31]
8	Chitosan-grafted graphene oxide (CS-GO) nanocomposite	2	96 (a)	[51]
9	Chitosan GO	3	310.4 (b)	[60]
10	Graphene oxide chitosan microspheres modified with α–FeO(OH)	3	97.69 (a)	[61]
11	Trioctylamine-exfoliated graphene oxide (TOA–EGO)	2.5–3	96.3	[23]
12	Graphene oxide (GO)	3	86.15 ± 0.8 (a)	Present study
13	GO-SiO_2	3	92.28 ± 0.45 (a)	Present study

It can be observed that an acidic pH in the range from 2.0 to 3.0 was found to be optimum in most of the studies. An acidic pH is often preferred for the removal of Cr(VI) from wastewater because it helps to convert the highly toxic and mobile Cr(VI) into its less toxic and less mobile Cr(III) form. This is because, at lower pH levels, the reduction of Cr(VI) to Cr(III) is more favorable and occurs more quickly. The reduction of Cr(VI) to Cr(III) occurs through various mechanisms, such as chemical reduction, microbial reduction, or electrochemical reduction. However, the efficiency of these mechanisms is affected by the pH of the solution. At a low pH, the chemical reduction of Cr(VI) is favored, while at higher pH values, microbial reduction is more favorable. Moreover, at lower pH values, the solubility of Cr(III) is significantly reduced, which leads to its precipitation and removal from the solution. This precipitation can occur in the form of various Cr(III) compounds, such as hydroxides, oxides, or salts. These precipitates are less soluble and less mobile than Cr(VI), reducing the risk of groundwater contamination and other adverse environmental impacts.

Compared with other adsorbents, a high absorbance percentage was recorded for synthesized GO and GO/SiO_2 at pH 3. Other adsorbents that are reported in the literature with a high removal percentage showed maximum adsorption at a pH < 3. A high surface area and the presence of oxygen-containing functional groups on the surface of GO facilitate the removal efficiency by complexation with Cr(VI) ions. The higher removal efficiency of GO/SiO_2 can be justified by its enhanced properties.

4. Conclusions

In this study, graphene oxide (GO) and graphene-oxide-coated silica nanoparticles were successfully synthesized and characterized by FTIR and Raman spectroscopy and TEM. The synthesized nanoparticles were used for the removal of chromium (VI) ions from water. The pH was optimized to 3, and the maximum adsorption percentage of Cr(VI) was observed at a concentration of 50 ppm for both of the adsorbents. A difference of 6.13% adsorption was observed between GO and GO-coated silica nanoparticles at optimum conditions. The results of the present study showed that both GO and GO-coated silica nanoparticles are effective adsorbents for the Cr(VI) removal from water. The adsorption of Cr(VI) on these synthesized adsorbents is strongly dependent on the pH values and initial concentration of the solution. GO-coated silica nanoparticles are excellent adsorbents compared to GO alone, and we highly recommend that they can be used in water treatment for removing Cr(VI). Further optimization studies are essential for the various applications of these nanoparticles.

Author Contributions: Conceptualization, K.P.J., S.T. and S.V.P.; methodology, K.P.J., S.A., S.K.S., S.T. and S.V.P.;validation, K.P.J., S.T., S.V.P. and S.A.; formal analysis, K.P.J., S.A., S.K.S., P.V., J.S.G. and S.V.P.; investigation, K.P.J., S.A., S.K.S., P.V., J.S.G. and S.V.P.; resources, K.P.J., S.T. and S.V.P.; data curation, K.P.J., S.T. and S.V.P.; writing—original draft preparation, K.P.J., S.A., S.K.S., P.V. and J.S.G.; writing—review and editing, K.P.J., S.T. and S.V.P.; supervision, S.T. and S.V.P.; project administration, S.T.; funding acquisition, S.T. All authors have read and agreed to the published version of the manuscript.

Funding: This work was supported by the DST Nanomission project (SR/NM/NT-1054/2015) by the Department of Science and Technology, Government of India, Defence Research and Development Organization (DRDO), Naval Research Board (NRB-429/MAT/18-19), RUSA scheme under Ministry of Education, Government of India, Mahatma Gandhi University for providing University JRF (4446/AcA6/2022/MGU) and Department of Science and Technology (DST) through Innovation of Science Pursuit for Inspire Research programme (INSPIRE-IF190284).

Data Availability Statement: Data are available upon request.

Acknowledgments: We would like to acknowledge Anu A.S. (TEM Engineer) at the International and Inter University Centre for Nanoscience and Nanotechnology, Mahatma Gandhi University, Kottayam, for performing the TEM Analysis.

Conflicts of Interest: On behalf of all authors, the corresponding author states that there is no conflict of interest.

References

1. Järup, L. Medical bulletin and undefined 2003: Hazards of heavy metal contamination. *Br. Med. Bull.* **2003**, *68*, 167–182. [CrossRef] [PubMed]
2. Jaishankar, M.; Tseten, T.; Anbalagan, N.; Mathew, B.B.; Beeregowda, K.N. Toxicity, mechanism and health effects of some heavy metals. *Interdiscip. Toxicol.* **2014**, *7*, 60. [CrossRef]
3. Brevik, E.C.; Burgess, L.C. *Soils and Human Health*; CRC Press: Boca Raton, FL, USA, 2012.
4. Sperling, M. Chromium. *Encycl. Anal. Sci. Second Ed.* **2004**, *113*. [CrossRef]
5. Rodríguez, M.C.; Barsanti, L.; Passarelli, V.; Evangelista, V.; Conforti, V.; Gualtieri, P. Effects of chromium on photosynthetic and photoreceptive apparatus of the alga *Chlamydomonas reinhardtii*. *Environ. Res.* **2007**, *105*, 234. [CrossRef]
6. Saha, R.; Nandi, R.; Saha, B. Chemistry, and undefined 2011: Sources and toxicity of hexavalent chromium. *J. Coord. Chem.* **2011**, *64*, 1782–1806. [CrossRef]
7. Li, Y.; Zhou, Q.; Ren, B.; Luo, J.; Yuan, J.; Ding, X.; Bian, H.; Yao, X. Trends and Health Risks of Dissolved Heavy Metal Pollution in Global River and Lake Water from 1970 to 2017. *Rev. Environ. Contam. Toxicol.* **2020**, *251*, 1–24.
8. Tumolo, M.; Ancona, V.; De Paola, D.; Losacco, D.; Campanale, C.; Massarelli, C.; Uricchio, V.F. Chromium pollution in European water, sources, health risk, and remediation strategies: An overview. *Int. J. Environ. Res. Public Health* **2020**, *17*, 5438. [CrossRef]
9. Vardhan, K.H.; Kumar, P.S.; Panda, R.C. A review on heavy metal pollution, toxicity and remedial measures: Current trends and future perspectives. *J. Mol. Liq.* **2019**, *290*, 111197. [CrossRef]
10. Lisiak-Zielińska, M.; Borowiak, K.; Budka, A.; Kanclerz, J.; Janicka, E.; Kaczor, A.; Żyromski, A.; Biniak-Pieróg, M.; Podawca, K.; Mleczek, M.; et al. How polluted are cities in central Europe?—Heavy metal contamination in Taraxacum officinale and soils collected from different land use areas of three representative cities. *Chemosphere* **2021**, *266*, 129113. [CrossRef]
11. Fu, F.; Wang, Q. Removal of heavy metal ions from wastewaters: A review. *J. Environ. Manag.* **2021**, *92*, 407. [CrossRef]
12. Singh, K.; Renu, N.A.; Agarwal, M. Methodologies for removal of heavy metal ions from wastewater: An overview. *Interdiscip. Environ. Rev.* **2017**, *18*, 124. [CrossRef]
13. Zhao, D.; Gao, X.; Chen, S.; Xie, F.; Feng, S.; Alsaedi, A.; Hayat, T.; Chen, C. Interaction between U(VI) with sulfhydryl groups functionalized graphene oxides investigated by batch and spectroscopic techniques. *J. Colloid Interface Sci.* **2018**, *524*, 129. [CrossRef] [PubMed]
14. Malik, L.A.; Bashir, A.; Qureashi, A.; Pandith, A.H. Detection and removal of heavy metal ions: A review. *Environ. Chem. Lett.* **2019**, *17*, 1495. [CrossRef]
15. Vakili, M.; Deng, S.; Cagnetta, G.; Wang, W.; Meng, P.; Liu, D.; Yu, G. Regeneration of chitosan-based adsorbents used in heavy metal adsorption: A review. *Sep. Purif. Technol.* **2019**, *224*, 373. [CrossRef]
16. Arora, R. Adsorption of heavy metals—A review. *Mater. Today Proc.* **2019**, *18*, 4745. [CrossRef]
17. Prachi; Gautam, P.; Madathil, D.; Nair, A.N.B. Nanotechnology in waste water treatment: A review. *Int. J. ChemTech Res.* **2013**, *5*, 2303.
18. Gangadhar, G.; Maheshwari, U.; Gupta, S. Application of nanomaterials for the removal of pollutants from effluent streams. *Nanosci. Nanotechnol.-Asia* **2012**, *2*, 140–150. [CrossRef]
19. Potts, J.R.; Dreyer, D.R.; Bielawski, C.W.; Ruoff, R.S. Graphene-based polymer nanocomposites. *Polymer* **2011**, *52*, 5–25. [CrossRef]
20. Sadegh, H.; Ali, G.A.; Gupta, V.K.; Makhlouf, A.S.H.; Shahryari-Ghoshekandi, R.; Nadagouda, M.N.; Sillanpää, M.; Megiel, E. The role of nanomaterials as effective adsorbents and their applications in wastewater treatment. *J. Nanostructure Chem.* **2017**, *7*, 1–14. [CrossRef]
21. Liu, X.; Ma, R.; Wang, X.; Ma, Y.; Yang, Y.; Zhuang, L.; Zhang, S.; Jehan, R.; Chen, J.; Wang, X. Graphene oxide-based materials for efficient removal of heavy metal ions from aqueous solution: A review. *Environ. Pollut.* **2019**, *252*, 62. [CrossRef]
22. Li, L.; Zhao, L.; Ma, J.; Tian, Y.; Tian, Y. Preparation of graphene oxide/chitosan complex and its adsorption properties for heavy metal ions. *Green Process. Synth.* **2020**, *9*, 294. [CrossRef]
23. Kumar, A.S.K.; Kakan, S.S.; Rajesh, N. A novel amine impregnated graphene oxide adsorbent for the removal of hexavalent chromium. *Chem. Eng. J.* **2013**, *230*, 328. [CrossRef]
24. Xu, L.; Wang, J. The application of graphene-based materials for the removal of heavy metals and radionuclides from water and wastewater. *Crit. Rev. Environ. Sci. Technol.* **2017**, *47*, 1042. [CrossRef]
25. Deng, J.H.; Zhang, X.R.; Zeng, G.M.; Gong, J.L.; Niu, Q.Y.; Liang, J. Simultaneous removal of Cd (II) and ionic dyes from aqueous solution using magnetic graphene oxide nanocomposite as an adsorbent. *Chem. Eng. J.* **2013**, *226*, 189–200. [CrossRef]
26. Peer, F.E.; Bahramifar, N.; Younesi, H. Removal of Cd (II), Pb (II) and Cu (II) ions from aqueous solution by polyamidoamine dendrimer grafted magnetic graphene oxide nanosheets. *J. Taiwan Inst. Chem. Eng.* **2018**, *87*, 225–240. [CrossRef]
27. Shahzad, A.; Miran, W.; Rasool, K.; Nawaz, M.; Jang, J.; Lim, S.R.; Lee, D.S. Heavy metals removal by EDTA-functionalized chitosan graphene oxide nanocomposites. *RSC Adv.* **2017**, *7*, 9764. [CrossRef]
28. Xu, J.; Cao, Z.; Zhang, Y.; Yuan, Z.; Lou, Z.; Xu, X.; Wang, X. A review of functionalized carbon nanotubes and graphene for heavy metal adsorption from water: Preparation, application, and mechanism. *Chemosphere* **2018**, *195*, 351–364. [CrossRef] [PubMed]

29. Manyangadze, M.; Chikuruwo, N.M.H.; Narsaiah, T.B.; Chakra, C.S.; Charis, G.; Danha, G.; Mamvura, T.A. Adsorption of lead ions from wastewater using nano silica spheres synthesized on calcium carbonate templates. *Heliyon* **2020**, *6*, e05309. [CrossRef]
30. Hummers, W.S., Jr.; Offeman, R.E. Preparation of graphitic oxide. *J. Am. Chem. Soc.* **1958**, *80*, 1339. [CrossRef]
31. Sheet, I.; Kabbani, A.; Holail, H. Removal of heavy metals using nanostructured graphite oxide, silica nanoparticles and silica/graphite oxide composite. *Energy Procedia* **2014**, *50*, 130. [CrossRef]
32. Mondal, N.K.; Chakraborty, S. Adsorption of Cr(VI) from aqueous solution on graphene oxide (GO) prepared from graphite: Equilibrium, kinetic and thermodynamic studies. *Appl. Water Sci.* **2020**, *10*, 1–10. [CrossRef]
33. Lee, D.W.; De Los Santos V, L.; Seo, J.W.; Felix, L.L.; Bustamante D, A.; Cole, J.M.; Barnes, C.H.W. The structure of graphite oxide: Investigation of its surface chemical groups. *J. Phys. Chem. B* **2010**, *114*, 5723–5728.
34. Li, D.; Müller, M.B.; Gilje, S.; Kaner, R.B.; Wallace, G.G. Processable aqueous dispersions of graphene nanosheets. *Nat. Nanotechnol.* **2008**, *3*, 101. [CrossRef] [PubMed]
35. Zhang, W.; Langmuir, H.C. Silica-graphene oxide hybrid composite particles and their electroresponsive characteristics. *ACS Publ.* **2012**, *28*, 7055. [CrossRef] [PubMed]
36. Björkman, Å. Thermische Klärschlammbehandlung. *Schweiz. Z. Hydrol.* **1969**, *31*, 632.
37. Eda, G.; Chhowalla, M. Chemically derived graphene oxide: Towards large-area thin-film electronics and optoelectronics. *Adv. Mater.* **2010**, *22*, 2392. [CrossRef]
38. Sun, L.; Yu, H.; Fugetsu, B. Graphene oxide adsorption enhanced by in situ reduction with sodium hydrosulfite to remove acridine orange from aqueous solution. *J. Hazard. Mater.* **2012**, *203–204*, 101. [CrossRef]
39. Kudin, K.N.; Ozbas, B.; Schniepp, H.C.; Prud, R.K.; Aksay, I.A.; Car, R. Raman Spectra of Graphite Oxide and Functionalized Graphene Sheets. *Nano Lett.* **2008**, *8*, 36–41.
40. Gu, J.; She, J.; Yue, Y. Micro/nanoscale thermal characterization based on spectroscopy techniques. *ES Energy Environ.* **2020**, *9*, 15–27. [CrossRef]
41. Khalili, D. Graphene oxide: A promising carbocatalyst for the regioselective thiocyanation of aromatic amines, phenols, anisols and enolizable ketones by hydrogen peroxide/KSCN in water. *N. J. Chem.* **2016**, *40*, 2547. [CrossRef]
42. Krishnamoorthy, K.; Veerapandian, M.; Yun, K.; Kim, S.J. The chemical and structural analysis of graphene oxide with different degrees of oxidation. *Carbon N. Y.* **2013**, *53*, 38. [CrossRef]
43. Mishima, K.; Du, X.; Sekiguchi, S.; Kano, N. Experimental and theoretical studies on the adsorption and desorption mechanisms of chromate ions on cross-linked chitosan. *J. Funct. Biomater.* **2017**, *8*, 3. [CrossRef]
44. Zhao, G.; Li, J.; Ren, X.; Chen, C.; Wang, X. Few-layered graphene oxide nanosheets as superior sorbents for heavy metal ion pollution management. *Environ. Sci. Technol.* **2011**, *45*, 10454. [CrossRef]
45. Shirkhanloo, H.; Khaligh, A.; Mousavi, H.Z.; Rashidi, A. Graphene oxide-packed micro-column solid-phase extraction combined with flame atomic absorption spectrometry for determination of lead (II) and nickel (II) in water. *J. Environ. Anal. Chem.* **2015**, *95*, 16–32. [CrossRef]
46. Zhang, L.-H.; Sun, Q.; Yang, C.; Lu, A.-H. Synthesis of magnetic hollow carbon nanospheres with superior microporosity for efficient adsorption of hexavalent chromium ions. *Sci. China Mater.* **2015**, *58*, 611. [CrossRef]
47. Hu, J.; Chen, C.; Zhu, X.; Wang, X. Removal of chromium from aqueous solution by using oxidized multiwalled carbon nanotubes. *J. Hazard. Mater.* **2009**, *162*, 1542. [CrossRef] [PubMed]
48. Matei, E.; Predescu, A.M.; Râpă, M.; Tarcea, C.; Pantilimon, C.M.; Favier, L.; Berbecaru, A.C.; Sohaciu, M.; Predescu, C. Removal of Chromium(VI) from Aqueous Solution Using a Novel Green Magnetic Nanoparticle–Chitosan Adsorbent. *Anal. Lett.* **2019**, *52*, 2416. [CrossRef]
49. Hu, J.; Lo, I.M.C.; Chen, G. Removal of Cr(VI) by magnetite nanoparticle. *Water Sci. Technol.* **2004**, *50*, 139. [CrossRef]
50. Liu, W.; Yang, L.; Xu, S.; Chen, Y.; Liu, B.; Li, Z.; Jiang, C. RSC Advances Efficient removal of hexavalent chromium from water by an adsorption-reduction mechanism with sandwiched nanocomposites. *RSC Adv.* **2018**, *8*, 15087–15093. [CrossRef]
51. Samuel, M.S.; Bhattacharya, J.; Raj, S.; Santhanam, N.; Singh, H.; Singh, N.D.P. Efficient removal of Chromium(VI) from aqueous solution using chitosan grafted graphene oxide (CS-GO) nanocomposite. *Int. J. Biol. Macromol.* **2019**, *121*, 285. [CrossRef] [PubMed]
52. Konicki, W.; Aleksandrzak, M.; Mijowska, E. Equilibrium and kinetics studies for the adsorption of Ni2+ and Fe3+ ions from aqueous solution by graphene oxide. *Polish J. Chem. Technol.* **2017**, *19*, 120. [CrossRef]
53. Donga, C.; Mishra, S.; Aziz, A.; Ndlovu, L.; Kuvarega, A.; Mishra, A.K. 3-Aminopropyl) Triethoxysilane (APTES) Functionalized Magnetic Nanosilica Graphene Oxide (MGO) Nanocomposite for the Comparative Adsorption of the Heavy Metal [Pb(II), Cd(II) and Ni(II)] Ions from Aqueous Solution. *J. Inorg. Organomet. Polym. Mater.* **2022**, *32*, 2235–2248. [CrossRef]
54. Ma, M.; Li, H.; Xiong, Y.; Dong, F. Rational design, synthesis, and application of silica/graphene-based nanocomposite: A review. *Mater. Des.* **2021**, *198*, 109367. [CrossRef]
55. Kalidhasan, S.; Kumar, A.S.K.; Rajesh, V.; Rajesh, N. Enhanced adsorption of hexavalent chromium arising out of an admirable interaction between a synthetic polymer and an ionic liquid. *Chem. Eng. J.* **2013**, *222*, 454. [CrossRef]
56. Kumar, A.S.K.; Rajesh, N.; Kalidhasan, S.; Rajesh, V. An enhanced adsorption methodology for the detoxification of chromium using n-octylamine impregnated Amberlite XAD-4 polymeric sorbent. *J. Environ. Sci. Health—Part A Toxic/Hazard. Subst. Environ. Eng.* **2011**, *46*, 1598. [CrossRef]

57. Zhang, L.; Song, F.; Wang, S.; Wang, H.; Yang, W.; Li, Y. Efficient removal of hexavalent chromium and congo red by graphene oxide/silica nanosheets with multistage pores. *ACS Publ.* **2020**, *65*, 4368. [CrossRef]
58. Hu, J.; Lo, I.M.C.; Chen, G. Fast removal and recovery of Cr(VI) using surface-modified jacobsite ($MnFe_2O_4$) nanoparticles. *Langmuir* **2005**, *21*, 11173. [CrossRef] [PubMed]
59. Hu, J.; Chen, G.; Lo, I.M.C. Removal and recovery of Cr(VI) from wastewater by maghemite nanoparticles. *Water Res.* **2005**, *39*, 4528. [CrossRef]
60. Najafabadi, H.H.; Irani, M.; Rad, L.R.; Haratameh, A.H.; Haririan, I. Removal of Cu2+, Pb2+ and Cr6+ from aqueous solutions using a chitosan/graphene oxide composite nanofibrous adsorbent. *RSC Adv.* **2015**, *5*, 16532. [CrossRef]
61. Liu, Y.; Shan, H.; Zeng, C.; Zhan, H.; Pang, Y. Removal of Cr(VI) from Wastewater Using Graphene Oxide Chitosan Microspheres Modified with α–FeO(OH). *Materials* **2022**, *15*, 4909. [CrossRef] [PubMed]

Disclaimer/Publisher's Note: The statements, opinions and data contained in all publications are solely those of the individual author(s) and contributor(s) and not of MDPI and/or the editor(s). MDPI and/or the editor(s) disclaim responsibility for any injury to people or property resulting from any ideas, methods, instructions or products referred to in the content.

crystals

MDPI

Article

Modelling the Structure and Optical Properties of Reduced Graphene Oxide Produced by Laser Ablation: Insights from XPS and Time-Dependent DFT

Igor V. Ershov [1,2,*], Anatoly A. Lavrentyev [3], Igor V. Bazhin [1], Olga M. Holodova [1], Natalia V. Prutsakova [1], Tatiana P. Zhdanova [1] and Dmitry L. Romanov [3]

1 Department of Physics, Don State Technical University, Rostov-on-Don 344000, Russia; ib_rnd@mail.ru (I.V.B.)
2 Institute of Emerging Technologies, Don State Technical University, Rostov-on-Don 344000, Russia
3 Department of Electrical Engineering and Electronics, Don State Technical University, Rostov-on-Don 344000, Russia; alavrentyev@donstu.ru (A.A.L.)
* Correspondence: iershov@donstu.ru

Abstract: Graphene derivatives such as reduced graphene oxide and graphene-based composites are regarded as highly promising materials for optoelectronics and photodetection applications. Recently, considerable interest has arisen in using facile and environmentally friendly methods of graphene production. Despite significant progress in experimental studies of graphene and graphene-based composites, some fundamental questions about their structures, and the interaction between components in these systems, remain open. In the present work, several atomistic models of oxidized graphene fragments have been proposed based on XPS compositional analysis and DFT calculations, representing reduced graphene oxide produced by laser irradiation. The composition of the oxygen-containing functional groups, their equilibrium configuration and influence on the electronic and optical properties of graphene sheets were determined. The nature of the low-lying excited states, as well as the photoactive regions, has also been studied for the proposed models. The calculated absorption spectra of the graphene sheets were compared with experimental UV-Vis spectrum of rGO produced by laser ablation.

Keywords: reduced graphene oxide; laser ablation; XPS; time-dependent DFT; Liouville–Lanczos approach

1. Introduction

From its discovery in 2004 to the present, graphene continues to attract great attention in various fields of scientific research and applications due to its unique electronic, thermal, mechanical and optical properties [1]. Graphene is an attractive material for optoelectronic and photodetection applications [2,3] because it offers a broad spectral bandwidth and fast response times. However, pure graphene exhibits zero bandgap and weak light absorption; in addition, it has low solubility in water and tends to aggregate due to interlayer π-π interactions, which limits its optoelectronic and some other potential applications. For example, the oxygen functional groups attached to the surfaces and edges of graphene sheet can significantly modify its electronic, mechanical and optical properties and allows graphene to be processed easily in dispersions, leading to the plethora of applications, including solar cells [4–7], sensors [8], supercapacitors [9], and many more. For this reason, most of the new promising composite materials are based on graphene derivatives such as graphene oxide (GO), reduced graphene oxide (rGO) [10,11] and zero-dimensional nanostructures composed of one or few layers [12]. The above materials have tunable properties and can form stable aqueous suspensions.

Recently GO considered only a simple and inexpensive material for the preparation of single and multilayer graphene by reduction, but the results of the latest research demonstrated a large number of unique optical and electronic properties of GO and rGO that means they should be regarded as independent nanomaterials with a wide range of applications [13–15]. Reduced graphene oxide is regarded as a highly promising material for optoelectronic applications. One of the widely discussed applications for rGO, due to the growing interest in flexible and stretchable electronic devices, is thin transparent protective and conductive films, which are mostly used in optoelectronic and photovoltaic devices as electrodes. Films with a thickness of several atomic layers could be easily deposited on any hydrophilic surface by simple drop-casting or spin coating methods [16]. Graphene derivatives are less expensive and more environmentally friendly in comparison with traditional ITO and can successfully replace it in most applications [10, 17]. Recently, silver- and gold-decorated rGO hybrid sheets were successfully fabricated and demonstrated low resistance at more than 83% transparency [17]. Ag–rGO films achieved a sheet conductivity two orders of magnitude higher than that of rGO films [17]. Flexible field-effect light-emitting diodes based on rGO were constructed by employing a graphene field-effect transistor [18]. To date, rGO has been successfully used in the fabrication of organic, perovskite, silicon and dye-sensitized solar cells and photodetectors [4–7,19]. Dye functionalized graphenes, such as rGO nanohybrids with porphyrins, were shown to possess great potential in the field of photocatalytic hydrogen production and optical sensing [20,21].

The common methods to obtain rGO include chemical and thermal reduction of GO. Recently, laser irradiation has been proposed as a promising tool for the reduction of GO [22–26]. In addition to the spatial control and energy efficiency, the laser reduction process is much cleaner and less toxic than any chemical method, as it does not require chemical reactions in acids, or long-term washing procedures [23,25]. Pulsed laser ablation in liquid has recently attracted increased attention for the synthesis of graphene and GO quantum dots [27,28]. In our previous work, we reported on graphenic oxidized films deposited on metal and dielectric substrates by pulsed laser ablation in an inert gas atmosphere [29]. We have found a significant reduction of oxygen-containing groups in comparison with the target material. The methods of graphene oxidation/reduction and fragmentation significantly affect its physical properties, including the mechanisms of conduction, light absorption and emission. Taking into account the diversity of the possible graphene derivatives, the application of theoretical methods that could correctly predict the spectroscopic properties of those materials can serve as a guidance for fabrication of such materials with desired properties. Generally, rGO is an amorphous system, and there is no direct experimental information that can provide a clear understanding of its local microscopic structure. At the same time, an understanding of the atomic structure can help reveal the mechanisms of interaction of rGO with other molecules, surfaces and nanostructures, and allow correct interpretation of spectroscopic data. To date there have been a number of papers on modelling graphene and graphene derivatives. There are two approaches in modelling graphene oxide. The first is based on periodic density functional theory calculations and assumes some degree of repeating order in oxygen atoms' distribution on the graphene surface [30–32]. The other approach consider graphene as a finite sheet decorated with oxygen-containing groups [21,33,34]. The finite models, in spite of the computational complexity, become more preferable as they are physically more reasonable. They allow reproduction of the edge functionalization of graphene and correctly describe the adsorption and interaction of graphene with molecules and different surfaces. Finite models have also been shown to reproduce the optical properties of graphene-based nanohybrid systems with good accuracy [21,34,35]. Among the various finite structural models proposed recently [36–38] the Lerf–Klinowski model (LK) [39] was explicitly or implicitly assumed in the majority of successful functionalization studies [38]. While the LK model is quite successful in describing the properties of graphene produced by "classical" chemical methods, it does not account for the structural features of reduced graphene oxide produced by physical methods, such as hydrothermal and laser reduction [25,40]. The successful validation of structural models relies on the combination of theoretical and experimental approaches. In this work, we present atomistic models of oxidized graphene sheets, produced by laser ablation, based on XPS compositional analysis and DFT calculations of electronic and optical properties.

2. Materials and Methods

The widely accepted Lerf–Klinowski model [39] and other finite structural models [36,37] describe GO as carbon planes heavily decorated with epoxy and hydroxyl groups on the surfaces and carbonyl and carboxyl groups at the edges of the nanosheets. These oxygen groups are mostly removed from the surface in rGO by the use of chemical reducing agents such as hydrazine and its derivatives, sodium borohydride or hydroiodic acid, or physical reducing agents such as heat and optical radiation. In the case of laser irradiation of GO or oxidized carbon material, several processes can occur including photochemical and photothermal removal of oxygen functional groups, which is accompanied by laser ablation, and structural reorganization of the newly formed carbon lattice into the planar sp^2-conjugated graphenic structure [26]. Fragmentation of carbon materials in solvents by laser irradiation can be explained within the model of Coulomb explosion [41]. As a result of this process in the laser plume, carbon precursors are ablated to the quantum size, and solvent can decompose into small molecules. The decomposed molecules can functionalize ablated graphenic fragments, which can significantly modify the final optoelectronic properties of the graphenic material. In contrast to chemically reduced graphene oxide, not only carboxyl and hydroxyl groups, but also ketones and quinones can remain on the edges of graphene sheets after laser ablation. Because of this, the Lerf–Klinowski model may not accurately describe the structure of graphene sheets produced by laser irradiation.

In this work, we propose several models of oxidized graphene, based on XPS compositional analysis of carbon films deposited on a silver substrate by laser ablation. The details of sample preparation and characterization are presented in supplementary information and in [29]. In our simulations, we considered rGO as a two-dimensional finite system, which contains randomly distributed graphene sheets of different sizes; recently this approach has been successfully applied to study GO/rGO-based nanohybrid systems [21,34,35]. XPS spectra allowed us to determine the type and amount of functional groups attached to carbon atoms, as well as the average carbon-to-oxygen ratio C/O in the graphene sheet. The core-level 1s carbon and oxygen spectra (Figure S2 in the Supplementary Materials) has shown that the ablated nanocarbon contains oxygen, singly bonded to aliphatic carbon (hydroxyl and/or epoxy groups), oxygen doubly bonded to aromatic and aliphatic carbon (carboxyl, carbonyl and quinone groups), and oxygen singly bonded to aromatic carbon (phenol groups) (Table S1 in the Supplementary Materials). The C/O ratio was estimated at ca. 6.6 which indicates that a large amount of the oxygen-containing groups were removed upon ablation. The partial restoration of the sp^2 honeycomb lattice was confirmed by Raman spectroscopy (Figure S3 in the Supplementary Materials). In this study we neglected the occurrence of point defects within the graphene sheets. Thus, in our models, graphene sheets consisted of a polyaromatic core with honeycomb lattice and oxygen functional groups at the edges, as illustrated in Figure 1.

Our approach was based on the search for structures with minimum total energy and a singlet ground-state electronic configuration. The characteristic dimensions of the sheets ranged from 1.6 nm to 2.3 nm, and the carbon-to-oxygen ratio within sheets ranged from 6.1 to 7. As a polyaromatic core, we considered stable structures with low and high symmetry (including D_{3h} and D_{6h}), containing both zigzag and armchair edges. For the final oxidized structures with the lowest ground state energy, the complete structural optimization was carried out using density functional theory within the plane-wave pseudopotential method [42].

The Kohn–Sham orbitals have been expanded in plane waves up to a kinetic energy cutoff of 60 Ry. The norm-conserving Hartwigsen–Goedeker–Hutter pseudopotentials were used. The oxidized sheets were separated from their periodic images in the supercell by at least 12 Å of vacuum. The atomic positions in the supercell were fully optimized with the force threshold of $2{\cdot}10^{-3}$ Ry/Å.

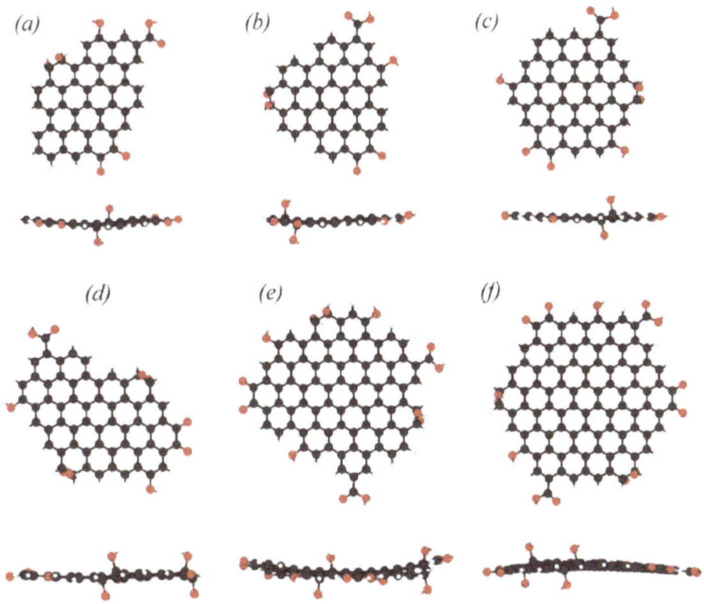

Figure 1. Atomistic models of oxidized graphene sheets: (**a**)—$C_{47}H_{18}O_7$, (**b**)—$C_{49}H_{18}O_7$, (**c**)—$C_{55}H_{18}O_8$, (**d**)—$C_{61}H_{22}O_{10}$, (**e**)—$C_{84}H_{26}O_{13}$, (**f**)—$C_{98}H_{24}O_{14}$.

The optical properties of the systems were studied using the linear-response time-dependent density functional theory. All computations were performed with the B3LYP hybrid functional. We compared the performance of different exchange-correlation functionals, including hybrid PBE0, BHLYP, and range-separated ωB97X functional. Semilocal GGA-functionals are known to underestimate excitation energies of simple polyaromatic molecules. However, it was found that hybrid functionals with different amounts of Hartree–Fock exchange give a rather large spread in the values of the excitation energies. The best agreement of calculated excitation energies with experimental absorption spectra of simple polyaromatic molecules was found using the B3LYP level of theory (Table S2 in the Supplementary Materials).

For each atomistic model, the first 25 singlet excited states corresponding to vertical electronic transitions were calculated. The excitation energies were determined by calculation the eigenvalues and eigenvectors of a Liouvillian superoperator using the iterative procedure analogous to Davidson diagonalization, as implemented in the turboTDDFT2.0 code [43]. Absorption spectra were calculated using the linear-response recursive Lanczos approach [44]. In this method the frequency-dependent dipole susceptibility is written as a matrix element of the resolvent of the system's Liouvillian. The imaginary part of the susceptibility has been plotted representing the absorption spectra.

3. Results and Discussion

Total energy calculations showed, that oxygen-containing functional groups in graphene sheets tend to be located on edge carbon atoms that do not belong to aromatic sextets (so-called K-regions). It has also been found that carbonyl groups are most stable when they form dimers on the K-regions, similar to the 1,2-benzoquinone structure. Calculations have shown that sp^3-bonded hydroxyl groups also tend to be located in the 1,2-configuration on the edge carbon atoms in K-regions, but are oriented on opposite sides of the rGO sheet. The rearrangement of phenolic and carboxyl groups on the edges did not significantly change the total energy of the systems under study.

The electronic structure and absorption spectra of the polyaromatic cores were calculated for a better understanding of the role of oxygen functional groups. The electronic structure and optical properties of polyaromatic molecules largely depend on their symmetry. Structures $C_{48}H_{18}$, $C_{54}H_{18}$ and $C_{96}H_{24}$ with D_{3h} and D_{6h} symmetry respectively exhibit degenerate HOMO/HOMO-1 and LUMO/LUMO + 1 orbitals leading to wider energy gaps and transitions at shorter wavelengths (Figure S4 in the Supplementary Materials). The highest energy gap of 3.11 eV was observed for the $C_{48}H_{18}$ molecule with D_{3h} symmetry (Table S3 in the Supplementary Materials).

Oxidation leads to symmetry reduction and splitting of degenerate levels. The ground state Kohn–Sham (KS) energy levels diagram of the oxidized graphene sheets is shown in Figure 2. All structures demonstrate singlet ground state electronic configuration and an energy gap in the range from 1.77 eV to 2.29 eV. Analysis of charge density distribution for the chosen molecular orbitals has shown that the highest occupied molecular orbitals of the systems are formed predominantly by π-electrons in an aromatic core, while the lowest unoccupied MOs are localized at the K-regions with carbonyl groups and originate from p-electrons of carbon and oxygen (Figure S5 in the Supplementary Materials).

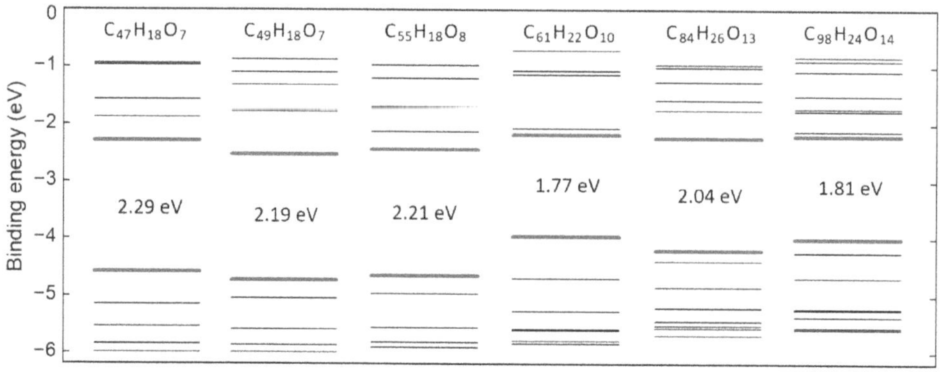

Figure 2. Ground state KS energy levels of rGO fragments.

Typically, large polycyclic aromatic molecules reveal three characteristic optical bands α, p, and β with increasing intensity in the absorption edge [45,46]. The α-band with the highest wavelength is usually very weak and symmetry-forbidden in high symmetry groups. The p-band originates from HOMO $\rightarrow$ LUMO transition, it can be observed in the low-symmetry structures $C_{46}H_{18}$, $C_{60}H_{20}$, and $C_{82}H_{24}$ as the first intense absorption peak (the S1 peak in Figure 3g,j,k). The β-band has the lowest wavelength of the three and corresponds to a transition from HOMO to the second lowest unoccupied orbital (LUMO + 1) [47]. In the high-symmetry polyaromatic cores $C_{48}H_{18}$, $C_{54}H_{18}$, $C_{96}H_{24}$ p-bands are symmetry-forbidden, and very intense β-bands represent the absorption edge (S3–S4 bands in Figure 3h,i,l). These structures exhibit narrow absorption spectra with only few transitions.

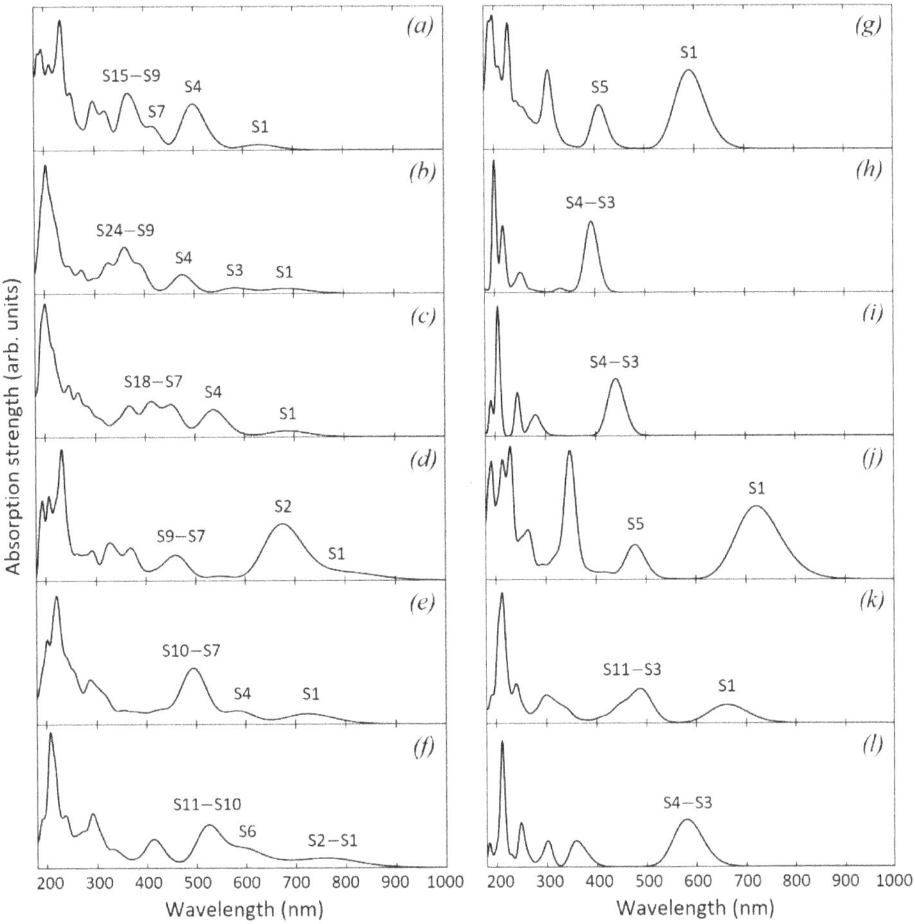

Figure 3. Calculated absorption spectra of oxidized graphene sheets and corresponding polyaromatic cores: (**a**)—$C_{47}H_{18}O_7$; (**b**)—$C_{49}H_{18}O_7$; (**c**)—$C_{55}H_{18}O_8$; (**d**)—$C_{61}H_{22}O_{10}$; (**e**)—$C_{84}H_{26}O_{13}$; (**f**)—$C_{98}H_{24}O_{14}$; (**g**)—$C_{46}H_{18}$; (**h**)—$C_{48}H_{18}$; (**i**)—$C_{54}H_{18}$; (**j**)—$C_{60}H_{20}$; (**k**)—$C_{82}H_{24}$; (**l**)—$C_{96}H_{24}$.

Oxygen functional groups modify the absorption spectra of unoxidized structures. Oxidized graphene sheets have the lowest C_1 symmetry, which leads to more allowed transitions and broader, less structured absorption spectra. For all oxidized structures, we can observe the occurrence of a long-wavelength absorption tail formed by the low-energy excited states (Figure 3a). We numbered the most characteristic absorption bands in the visible range in accordance with those excited states that make the dominant contribution at corresponding frequencies. Thus, for example, in the model $C_{47}H_{18}O_7$, the absorption edge is formed due to $S_0 \rightarrow S_1$ transition (Figure 3a), and the second intense absorption peak (denoted S4) at 500 nm is due to $S_0 \rightarrow S_4$ (it originates from the S_1 transition of the polyaromatic core $C_{46}H_{18}$ (Figure 3g). The data on excitation energies, wavelengths, oscillator strengths and transition coefficients for the considered models are presented in Table 1.

Table 1. Compound, the most intense singlet–singlet transition, photon energy, wavelength, sum of diagonal elements of the imaginary part of dynamic polarizability and principal transition components in terms of ground-state KS orbitals (H = HOMO; L = LUMO).

Compound	S → S Transition	E (eV)	λ (nm)	$\Sigma\chi_{ii}$ (e^2a^2/eV)	Transition Components
$C_{47}H_{18}O_7$	$S_0 \rightarrow S_1$	1.95	636	0.08	H → L (99%)
	$S_0 \rightarrow S_4$	2.48	500	0.43	H → L + 1 (58%) H → L + 2 (18%)
	$S_0 \rightarrow S_7$	2.94	422	0.20	H-1 → L + 1 (53%) H → L + 2 (25%)
	$S_0 \rightarrow S_{13}$	3.39	366	0.29	H-5 → L (75%)
$C_{49}H_{18}O_7$	$S_0 \rightarrow S_1$	1.85	670	0.18	H → L (95%)
	$S_0 \rightarrow S_3$	2.16	574	0.17	H-1 → L (92%)
	$S_0 \rightarrow S_4$	2.60	477	0.30	H → L + 1 (72%)
	$S_0 \rightarrow S_9$	3.20	387	0.40	H → L + 2 (40%) H-1 → L + 1 (21%)
	$S_0 \rightarrow S_{15}$	3.46	358	0.17	H-1 → L + 2 (34%) H-2 → L + 1 (17%)
$C_{55}H_{18}O_8$	$S_0 \rightarrow S_1$	1.84	674	0.12	H → L (92%)
	$S_0 \rightarrow S_4$	2.33	532	0.52	H → L + 1 (44%) H-1 → L (40%)
	$S_0 \rightarrow S_7$	2.76	449	0.51	H → L + 2 (55%) H-1 → L + 1 (21%)
	$S_0 \rightarrow S_9$	2.98	416	0.29	H-1 → L + 2 (45%) H → L + 3 (40%)
	$S_0 \rightarrow S_{11}$	3.03	409	0.22	H → L + 3 (46%) H-1 → L + 2 (29%)
$C_{61}H_{22}O_{10}$	$S_0 \rightarrow S_1$	1.58	785	0.18	H → L (85%)
	$S_0 \rightarrow S_2$	1.81	685	1.43	H → L + 1 (84%)
	$S_0 \rightarrow S_7$	2.59	479	0.17	H → L + 2 (51%) H-1 → L + 1 (21%)
	$S_0 \rightarrow S_9$	2.79	444	0.24	H-2 → L (78%)
$C_{84}H_{26}O_{13}$	$S_0 \rightarrow S_1$	1.74	713	0.41	H → L (93%)
	$S_0 \rightarrow S_4$	2.18	569	0.28	H → L + 1 (66%) H-1 → L + 2 (10%)
	$S_0 \rightarrow S_7$	2.50	496	1.30	H-1 → L + 1 (41%) H → L + 2 (29%)
	$S_0 \rightarrow S_{10}$	2.63	471	0.80	H-1 → L + 2 (33%) H → L + 3 (26%) H-3 → L (26%)
$C_{98}H_{24}O_{14}$	$S_0 \rightarrow S_1 + S_2$	1.61–1.63	761–770	0.43	H → L (61%) H → L + 1 (27%)
	$S_0 \rightarrow S_6$	2.10	590	0.72	H → L + 2 (48%) H → L + 3 (18%)
	$S_0 \rightarrow S_{11}$	2.37	523	1.09	H-1 → L + 3 (48%) H-2 → L + 1 (13%)

In order to better understand the individual absorption bands and photo-active regions we calculated the real-space electron charge-density response to a perturbing electric field $n'(\mathbf{r},\omega)$ [48]:

$$n'(r,\omega) = 2\sum_{v=1}^{N_v} \varphi_v^o(r)\left(\varphi_v'(r,\omega) + \varphi_v'^*(r,-\omega)\right), \quad (1)$$

where $\varphi_v^o(r)$-unperturbed (ground-state) KS orbitals, N_v-number of occupied KS states, $\varphi_v'(r,\omega) = \varphi_v(r,\omega) - \varphi_v^o(r)$-first-order response orbitals at frequency ω. The factor 2 accounts for the spin degeneracy of molecular orbitals in non-magnetic systems.

Response charge density of the most prominent transitions in the oxidized graphene sheets is shown in Figure 4. So, for example, in the case of the $C_{47}H_{18}O_7$ model, it can be seen that the weak S1 absorption band appears due to an electron-density response associated with electron-withdrawing carbonyl groups and neighboring carbon atoms. In contrast, the intense S4 band in this model, which has a maximum at 500 nm, is formed due to the response of the electron density of carbon atoms in the polyaromatic core. This density is delocalized throughout the graphene sheet and is not related to the activity of oxygen groups. Therefore, the S4 band originates from the $S_0 \rightarrow S_1$ transition in the $C_{46}H_{18}$ core. Transitions in the short-wave visible and near-UV regions are of mixed origin. Thus, for example, the absorption peak at 378 nm in the $C_{47}H_{18}O_7$ structure is due to the response of the electron density localized both on the carbon atoms of the polyaromatic core and on the hydroxyl, carboxyl, and carbonyl groups.

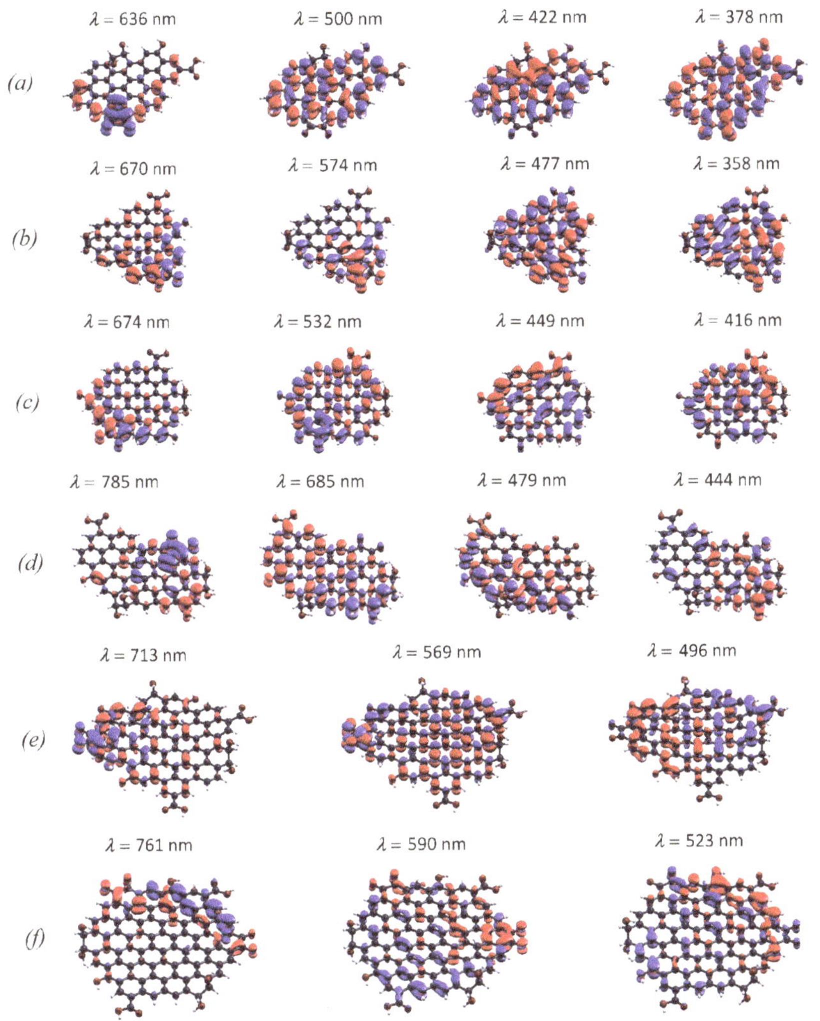

Figure 4. Response charge density to a perturbing electric field of selected wavelength for the oxidized graphene models: (**a**)—$C_{47}H_{18}O_7$; (**b**)—$C_{49}H_{18}O_7$; (**c**)—$C_{55}H_{18}O_8$; (**d**)—$C_{61}H_{22}O_{10}$; (**e**)—$C_{84}H_{26}O_{13}$; (**f**)—$C_{98}H_{24}O_{14}$.

The most significant changes in absorption spectra were found for the structures $C_{49}H_{18}O_7$ and $C_{55}H_{18}O_8$ (Figure 3b,c). While the $C_{48}H_{18}$ structure exhibits absorption only in the ultraviolet region, the optical spectrum of the structure $C_{49}H_{18}O_7$ covers the entire visible range. The $C_{49}H_{18}O_7$ and $C_{55}H_{18}O_8$ models have a similar absorption spectrum. In the long-wave range, according to Figure 3 and Table 1, the absorption spectrum of these systems is formed due to the transitions between two highest occupied and two lowest unoccupied molecular orbitals. In both structures the S1 absorption edge corresponds to the transition from the ground state to the first singlet excited state. It can be seen from the analysis of transition components that the $S_0 \rightarrow S_1$ excitation is entirely determined by the HOMO → LUMO transition in these models. From Figure 4b,c one can see that the absorption edge is related to the optical activity of the carbonyl groups and neighboring carbon atoms, similar to the $C_{47}H_{18}O_7$ model. In the $C_{49}H_{18}O_7$ model, the S3 absorption band at 574 nm is also related to this photo-active region because it is formed mainly by the transition HOMO-1 → LUMO to the lowest unoccupied orbital. According to Figure 4b,c the S4 absorption band in the models $C_{49}H_{18}O_7$ and $C_{55}H_{18}O_8$ is largely due to the optical response of carboxylic and carbonyl functional groups. In the case of the $C_{61}H_{22}O_{10}$ model the low-symmetry polyaromatic core produces an intense S2 absorption band associated with the low-energy $\pi \rightarrow \pi^*$ transitions (HOMO → LUMO + 1). The spectrum also features the weak absorption edge at 785 nm due to the response of carbonyl groups as in the previous models (Figure 4d). The models $C_{84}H_{26}O_{13}$ and $C_{98}H_{24}O_{14}$ with the largest number of carbon atoms in the core demonstrate absorption spectra similar in structure despite differences in symmetry and C/O ratio. The long-wavelength absorption tail as in the previous models is formed by the HOMO → LUMO and HOMO → LUMO + 1 transitions (Figure 3e,f) and is related to the response of the electron-withdrawing oxygen functional groups and response of polyaromatic core (Figure 4e,f). The most intense absorption bands in these models are observed in the wavelength range of 500–600 nm. In the system $C_{84}H_{26}O_{13}$ the absorption maximum in the visible region is at 496 nm and is related to the electronic transitions in the sp^2-carbon network. In the $C_{98}H_{24}O_{14}$ model, the absorption maximum is at 523 nm, and is due to the response of the zigzag edges as well as phenolic and carbonyl functional groups.

In order to understand which models better reproduce the optical properties of rGO we made a comparison of the calculated absorption spectra with the experimental UV-Vis spectrum of rGO suspensions, produced by laser ablation of high purity carbon tape (Figure 5). It should be noted that rGO samples measured in the experiment consist of multiple graphene sheets with different size, symmetry and arrangement of functional groups, thus the experimental absorption spectrum yields an averaged picture, and a single finite model is unable to reproduce this. The experimental spectrum features the absorption maximum at 240 nm which corresponds to a π-π^* plasmon peak, and a relatively continuous absorption tail up to 1000 nm. There are also no distinguishable features corresponding to n-π^* transitions.

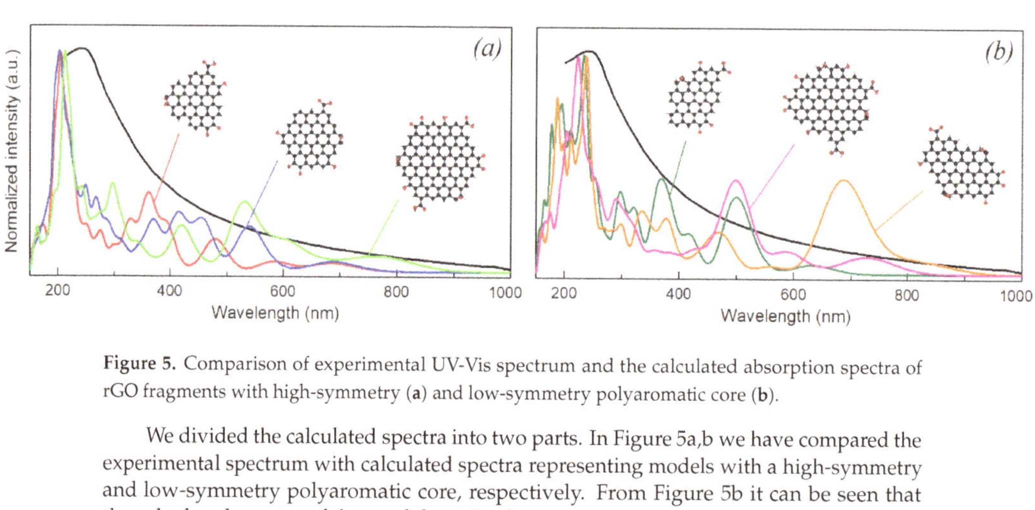

Figure 5. Comparison of experimental UV-Vis spectrum and the calculated absorption spectra of rGO fragments with high-symmetry (**a**) and low-symmetry polyaromatic core (**b**).

We divided the calculated spectra into two parts. In Figure 5a,b we have compared the experimental spectrum with calculated spectra representing models with a high-symmetry and low-symmetry polyaromatic core, respectively. From Figure 5b it can be seen that the calculated spectra of the models with a low-symmetry polyaromatic core accurately reproduce the position of the UV absorption maximum, but feature intense absorption peaks in the visible region, which are not observed in the experiment. In contrast, the spectra of the high-symmetry models are shown to better reproduce the experimental absorption tail throughout the visible range, though the absorption maximum was found to be blue-shifted in these models. The best agreement with experiment in the visible range was found for the combination of models $C_{47}H_{18}O_7$, $C_{49}H_{18}O_7$ and $C_{55}H_{18}O_8$. The calculated coefficient of determination for the weighted sum of these models in the wavelength range from 380 to 800 nm was 0.91.

4. Conclusions

In the present work, we proposed several atomistic models of rGO fragments based on XPS composition analysis of graphenic carbon produced by pulsed laser ablation and studied their optical properties using time-dependent DFT. It was shown that carbonyl and sp^3-bonded hydroxyl functional groups in graphene sheets tend to locate on edge carbon atoms that do not belong to aromatic sextets; in addition, carbonyl groups are most stable when they form dimers.

Optical absorption of rGO, produced by laser reduction in the long-wavelength region, may be due to the optical activity of carbonyl groups, and low-energy $\pi \rightarrow \pi^*$ transitions in the polyaromatic core of graphene sheets. The symmetry of the polyaromatic core affects the absorption intensity in the visible region, and the position of the absorption maximum in the UV region. A low-symmetry polyaromatic core in oxidized graphene sheet produces intense absorption bands in the visible range, which originate from intense p-bands. These bands were found to be blue-shifted and had a lower intensity when compared to the original p-bands. The calculated absorption spectra of the low-symmetry models were shown to accurately reproduce the experimental position of the $\pi \rightarrow \pi^*$ plasmon peak in the UV region, while the high-symmetry models were better at reproducing the absorbance in the visible region.

Understanding the atomic structure of graphene derivatives and its correlation with electronic and optical properties can help in the interpretation of new spectroscopic data and can serve as a guidance for the fabrication of new graphene-based hybrid materials and interfaces for optoelectronic applications.

Supplementary Materials: The following supporting information can be downloaded at: https://www.mdpi.com/article/10.3390/cryst13040600/s1, Figure S1: Survey XPS spectrum and carbon 1s deconvoluted core-level spectrum of target material; Figure S2: Survey XPS spectrum of rGO deposited on silver substrate (top) and 1s deconvoluted core-level spectra of carbon and oxygen (bottom) Table S1: XPS composition analysis, fitting parameters and C/O ratio of the as-prepared sample; Table S2: Calculated and experimental positions of p- and β-bands in the absorption spectra of simple nanographenes. Calculations performed using plane-wave pseudopotential method and B3LYP XC functional. Figure S3: Raman spectrum of the as-prepared sample; Figure S4: Ground state KS energy levels of polyaromatic cores; Table S3: Composition, C/O ratio in graphene sheet, positions of HOMO and LUMO, and energy gap of the rGO sheets and corresponding aromatic cores, calculated using plane-wave pseudopotential approach and B3LYP functional; Figure S5: Charge density distribution in real space for the frontier molecular orbitals: HOMO, LUMO, HOMO-1, LUMO + 1. (**a**)-$C_{47}H_{18}O_7$, (**b**)-$C_{49}H_{18}O_7$, (**c**)-$C_{55}H_{18}O_8$, (**d**)-$C_{61}H_{22}O_{10}$, (**e**)-$C_{84}H_{26}O_{13}$, (**f**)-$C_{98}H_{24}O_{14}$.

Author Contributions: Conceptualization, I.V.E. and A.A.L.; Data curation, A.A.L. and O.M.H.; Formal analysis, A.A.L.; Investigation, I.V.E., I.V.B. and N.V.P.; Methodology, I.V.E. and A.A.L.; Project administration, A.A.L.; Resources, I.V.E.; Software, I.V.B., D.L.R. and T.P.Z.; Supervision, I.V.E.; Validation, N.V.P. and I.V.B.; Visualization, O.M.H.; Writing-original draft, I.V.E.; Writing-review & editing, A.A.L. All authors have read and agreed to the published version of the manuscript.

Funding: This research received no external funding.

Institutional Review Board Statement: Not applicable.

Informed Consent Statement: Not applicable.

Data Availability Statement: Not applicable.

Acknowledgments: Authors would like to acknowledge I.N. Efremenko and E.A. Portnova for the assistance in organizing high performance computations at the Don Engineering Centre of DSTU.

Conflicts of Interest: The authors declare no conflict of interest.

References

1. Geim, A.K. Graphene: Status and prospects. *Science* **2009**, *324*, 1530–1534. [CrossRef]
2. Bonaccorso, F.; Sun, Z.; Hasan, T.; Ferrari, A.C. Graphene photonics and optoelectronics. *Nat. Photon.* **2010**, *4*, 611–622. [CrossRef]
3. Mueller, T.; Xia, F.; Avouris, P. Graphene photodetectors for high-speed optical communications. *Nat. Photon.* **2010**, *4*, 297–301. [CrossRef]
4. Mahakul, P.C.; Subramaniam, B.V.R.S.; Mahanandia, P. Mesoscopic investigation of the effect of MWCNT/rGO network on the performance of P3HT:PC60BM solar cells. *Mater. Chem. Phys.* **2019**, *226*, 113–117. [CrossRef]
5. Liu, R.; Qiu, R.; Zou, T.; Liu, C.; Chen, J.; Dai, Q.; Zhang, S.; Zhou, H. SnO2-rGO nanocomposite as an efficient electron transport layer for stable perovskite solar cells on AZO substrate. *Nanotechnology* **2018**, *30*, 075202. [CrossRef]
6. Jiang., X.; Wang, Z.; Han, W.; Liu, Q.; Lu, S.; Wen, Y.; Hou, J.; Huang, F.; Peng, S.; He, D.; et al. High performance silicon–organic hybrid solar cells via improving conductivity of PEDOT:PSS with reduced graphene oxide. *Appl. Surf. Sci.* **2017**, *407*, 398–404. [CrossRef]
7. Gopi, C.V.V.M.; Singh, S.; Reddy, A.E.; Kim, H.J. CNT@rGO@MoCuSe composite as an efficient counter electrode for quantum dot-sensitized solar cells. *ACS Appl. Mater. Interfaces* **2018**, *10*, 10036–10042. [CrossRef]
8. Liu, J.; Xu, M.; Wang, B.; Zhou, Z.; Wang, L. Fluorescence sensor for detecting protamines based on competitive interactions of polyacrylic acid modified with sodium 4-amino-1-naphthalenesulfonate with protamines and aminated graphene oxide. *RSC Adv.* **2017**, *7*, 1432–1438. [CrossRef]
9. Li, Z.; Gadipelli, S.; Yang, Y.; Guo, Z. Design of 3D Graphene-Oxide Spheres and Their Derived Hierarchical Porous Structures for High Performance Supercapacitors. *Small* **2017**, *13*, 1702474. [CrossRef]
10. Dideikin, A.T.; Vul', A.Y. Graphene oxide and derivatives: The place in graphene family. *Front. Phys.* **2019**, *6*, 149. [CrossRef]
11. Tarcan, R.; Todor-Boer, O.; Petrovai, I.; Leordean, C.; Astilean, S.; Botiz, I. Reduced graphene oxide today. *J. Mater. Chem. C* **2020**, *8*, 1198–1224. [CrossRef]
12. Bacon, M.; Bradley, S.J.; Nann, T. Graphene quantum dots. *Part. Part. Syst. Charact.* **2014**, *31*, 415–428. [CrossRef]
13. Novoselov, K.S.; Fal'ko, V.I.; Colombo, L.; Gellert, P.R.; Schwab, M.G.; Kim, K. A roadmap for graphene. *Nature* **2012**, *490*, 192–200. [CrossRef]
14. Kakavelakis, G.; Maksudov, T.; Konios, D.; Paradisanos, I.; Kioseoglou, G.; Stratakis, E.; Kymakis, E. Efficient and highly air stable planar inverted perovskite solar cells with reduced graphene oxide doped PCBM electron transporting layer. *Adv. Energy Mater.* **2017**, *7*, 1602120. [CrossRef]
15. Naderi, H.R.; Sobhani-Nasab, A.; Rahimi-Nasrabadi, M.; Ganjali, M.R. Decoration of nitrogen-doped reduced graphene oxide with cobalt tungstate nanoparticles for use in high-performance supercapacitors. *Appl. Surf. Sci.* **2017**, *423*, 1025–1034. [CrossRef]
16. Yamaguchi, H.; Eda, G.; Mattevi, C.; Kim, H.K.; Chhowalla, M. Highly uniform 300 mm wafer-Scale deposition of single and multilayered chemically derived graphene thin films. *ACS Nano* **2010**, *4*, 524–528. [CrossRef] [PubMed]
17. He, L.; Tjong, S.C. Silver-decorated reduced graphene oxides as novel building blocks for transparent conductive films. *RSC Adv.* **2017**, *7*, 2058–2065. [CrossRef]
18. Wang, X.; Tian, H.; Mohammad, M.A.; Li, C.; Wu, C.; Yang, Y.; Ren, T.L. A spectrally tunable all-graphene-based flexible field-effect light-emitting device. *Nat. Commun.* **2015**, *6*, 7767. [CrossRef]
19. Yu, X.X.; Yin, H.; Li, H.-X.; Zhao, H.; Li, C.; Zhu, M.-Q. A novel high-performance self-powered UV-vis-NIR photodetector based on a CdS nanorod array/reduced graphene oxide film heterojunction and its piezo-phototronic regulation. *J. Mater. Chem. C* **2018**, *6*, 630–636. [CrossRef]
20. Xu, Y.; Zhao, L.; Bai, H.; Hong, W.; Li, C.; Shi, G. Chemically converted graphene induced molecular flattening of 5,10,15,20-Tetrakis(1-methyl-4-pyridinio)porphyrin and its application for optical detection of cadmium(II) ions. *J. Am. Chem. Soc.* **2009**, *131*, 13490–13497. [CrossRef]
21. Siklitskaya, A.; Gacka, E.; Larowska, D.; Mazurkiewicz-Pawlicka, M.; Malolepszy, A.; Stobiński, L.; Marciniak, B.; Lewandowska-Andrałojć, A.; Kubas, A. Lerf–Klinowski-type models of graphene oxide and reduced graphene oxide are robust in analyzing non-covalent functionalization with porphyrins. *Sci. Rep.* **2021**, *11*, 7977. [CrossRef] [PubMed]
22. Kymakis, E.; Petridis, C.; Anthopoulos, T.D.; Stratakis., E. Laser-assisted reduction of graphene oxide for flexible, large-area optoelectronics. *IEEE J. Sel. Top. Quant. Electr.* **2014**, *20*, 106–115. [CrossRef]
23. Bhattacharjya, D.; Kim, C.-H.; Kim, J.-H.; You, I.-K.; In, J.B.; Lee, S.-M. Fast and controllable reduction of graphene oxide by low-cost CO_2 laser for supercapacitor application. *Appl. Surf. Sci.* **2018**, *462*, 353–361. [CrossRef]
24. Evlashin, S.; Dyakonov, P.; Khmelnitsky, R.; Dagesyan, S.; Klokov, A.; Sharkov, A.; Timashev, P.; Minaeva, S.; Maslakov, K.; Svyakhovskiy, S.; et al. Controllable Laser Reduction of Graphene Oxide Films for Photoelectronic Applications. *ACS Appl. Mater. Interfaces* **2016**, *8*, 28880–28887. [CrossRef] [PubMed]

25. Rodriguez, R.D.; Khalelov, A.; Postnikov, P.S.; Lipovka, A.; Dorozhko, E.; Amin, I.; Murastov, G.V.; Chen, J.-J.; Sheng, W.; Trusova, M.E.; et al. Beyond graphene oxide: Laser engineering functionalized graphene for flexible electronics. *Mater. Horiz.* **2020**, *7*, 1030–1041. [CrossRef]
26. Zhao, Y.; Han, Q.; Cheng, Z.; Jiang, L.; Qu, L. Integrated graphene systems by laser irradiation for advanced devices. *Nano Today* **2017**, *12*, 14–30. [CrossRef]
27. Donate-Buendia, C.; Torres-Mendieta, R.; Pyatenko, A.; Falomir, E.; Fernández-Alonso, M.; Mínguez-Vega, G. Fabrication by laser irradiation in a continuous flow jet of carbon quantum dots for fluorescence imaging. *ACS Omega* **2018**, *3*, 2735–2742. [CrossRef]
28. Kang, S.; Ryu, J.H.; Lee, B.; Jung, K.H.; Shim, K.B.; Han, H.; Kim., K.M. Laser wavelength modulated pulsed laser ablation for selective and efficient production of graphene quantum dots. *RSC Adv.* **2019**, *9*, 13658–13663. [CrossRef]
29. Ershov, I.V.; Lavrentyev, A.A.; Prutsakova, N.V.; Holodova, O.M.; Mardasova, I.V.; Zhdanova, T.P.; Kozakov, A.T. Characterization of graphenic carbon produced by pulsed laser ablation of sacrificial carbon tapes. *Appl. Sci.* **2021**, *11*, 11972. [CrossRef]
30. Boukhvalov, D.W.; Katsnelson, M.I. Modeling of Graphite Oxide. *J. Am. Chem. Soc.* **2008**, *130*, 10697–10701. [CrossRef]
31. Lahaye, R.J.W.E.; Jeong, H.K.; Park, C.Y.; Lee, Y.H. Density functional theory study of graphite oxide for different oxidation levels. *Phys. Rev. B* **2009**, *79*, 125435. [CrossRef]
32. Lundie, M.; Šljivančanin, Z.; Tomić., S. Electronic and optical properties of reduced graphene oxide. *J. Mater. Chem. C* **2015**, *3*, 7632–7641. [CrossRef]
33. Hernández Rosas, J.J.; Ramírez Gutiérrez, R.E.; Escobedo-Morales, A.; Chigo Anota, E. First principles calculations of the electronic and chemical properties of graphene, graphane, and graphene oxide. *J. Mol. Model.* **2011**, *17*, 1133–1139. [CrossRef] [PubMed]
34. Gacka, E.; Burdzinski, G.; Marciniak, B.; Kubas, A. Interaction of light with a non-covalent zinc porphyrin–graphene oxide nanohybrid. *Phys. Chem. Chem. Phys.* **2020**, *22*, 13456–13466. [CrossRef] [PubMed]
35. Gacka, E.; Wojcik, A.; Mazurkiewicz-Pawlicka, M.; Malolepszy, A.; Stobiński, L.; Kubas, A.; Hug, G.L.; Marciniak, B.; Lewandowska-Andralojc. A. Noncovalent Porphyrin–Graphene Oxide Nanohybrids: The pH-Dependent Behavior. *J. Phys. Chem. C* **2019**, *123*, 3368–3380. [CrossRef]
36. Szabo, T.; Berkesi, O.; Forgo, P.; Josepovits, K.; Sanakis, Y.; Petridis, D.; Dekany, I. Evolution of Surface Functional Groups in a Series of Progressively Oxidized Graphite Oxides. *Chem. Mater.* **2006**, *18*, 2740–2749. [CrossRef]
37. Gao, W.; Alemany, L.B.; Ci, L.; Ajayan, P.M. New insights into the structure and reduction of graphite oxide. *Nature Chem.* **2009**, *1*, 403–408. [CrossRef]
38. Dreyer, D.R.; Todd, A.D.; Bielawski., C.W. Harnessing the chemistry of graphene oxide. *Chem. Soc. Rev.* **2014**, *43*, 5288–5301. [CrossRef]
39. Lerf, A.; He, H.; Forster, M.; Klinowski, J. Structure of graphite oxide revisited. *J. Phys. Chem. B* **1998**, *102*, 4477–4482. [CrossRef]
40. Diez, N.; Sliwak, A.; Gryglewicz, S.; Grzyb, B.; Gryglewicz, G. Enhanced reduction of graphene oxide by high-pressure hydrothermal treatment. *RSC Adv.* **2015**, *5*, 81831. [CrossRef]
41. Zhang, D.; Gokce, B.; Barcikowski, S. Laser Synthesis and Processing of Colloids: Fundamentals and Applications. *Chem. Rev.* **2017**, *117*, 3990–4103. [CrossRef] [PubMed]
42. Giannozzi, P.; Andreussi, O.; Brumme, T.; Bunau, O.; Nardelli, M.B.; Calandra, M.; Car, R.; Cavazzoni, C.; Ceresoli, D.; Cococcioni, M.; et al. Advanced capabilities for materials modelling with Quantum ESPRESSO. *J. Phys. Condens. Matter.* **2017**, *29*, 465901. [CrossRef] [PubMed]
43. Ge, X.; Binnie, S.J.; Rocca, D.; Gebauer, R.; Baroni, S. turboTDDFT 2.0—Hybrid functionals and new algorithms within time-dependent density-functional perturbation theory. *Comp. Phys. Comm.* **2014**, *185*, 2080–2089. [CrossRef]
44. Gruning, M.; Marini, A.; Gonze, X. Implementation and testing of Lanczos-based algorithms for Random-Phase Approximation eigenproblems. *Comput. Mater. Sci.* **2011**, *50*, 2148–2156. [CrossRef]
45. Clar, E. *Polycyclic Hydrocarbons*; Academic Press and Springer-Verlag: London, UK, 1964; Volume 1–2.
46. Rieger, R.; Mullen, K. Forever young: Polycyclic aromatic hydrocarbons as model cases for structural and optical studies. *J. Phys. Org. Chem.* **2010**, *23*, 315–325. [CrossRef]
47. Fetzer, J.C. *Large (C >= 24) Polycyclic Aromatic Hydrocarbons*; John Wiley & Sons: New York, NY, USA, 2000.
48. Malcioglu, O.B.; Gebauer, R.; Rocca, D.; Baroni, S. turboTDDFT—A code for the simulation of molecular spectra using the Liouville–Lanczos approach to time-dependent density-functional perturbation theory. *Comp. Phys. Commun.* **2011**, *182*, 1744–1754. [CrossRef]

Disclaimer/Publisher's Note: The statements, opinions and data contained in all publications are solely those of the individual author(s) and contributor(s) and not of MDPI and/or the editor(s). MDPI and/or the editor(s) disclaim responsibility for any injury to people or property resulting from any ideas, methods, instructions or products referred to in the content.

crystals

MDPI

Article

Double-Heterostructure Resonant Tunneling Transistors of Surface-Functionalized Sb and Bi Monolayer Nanoribbons

Qi Gong and Guiling Zhang *

School of Material Science and Chemical Engineering, Harbin University of Science and Technology, Harbin 150040, China

* Correspondence: guiling-002@163.com

Abstract: Zigzag nanoribbons tailored from chemically surface-modified Sb or Bi monolayers by methyl, amino or hydroxyl are investigated through first-principles electronic-structure calculations to explore their potential applications in topological transport nanoelectronics. It is verified by Dirac-point-like energy dispersion of band-edges near Fermi level that the scattering-forbidden edge-states of these nanoribbons can give a topological conductive channel with extremely high electron mobility. Accordingly, Sb/$SbXH_n$/Sb and Bi/$BiXH_n$/Bi nanoribbon double-heterostructures ($SbXH_n$ or $BiXH_n$: XH_n = CH_3, NH_2, OH) are designed as resonant tunneling transistors and modeled by bipolar transport devices with their electron transport characteristics being calculated by nonequilibrium Green's function combined first-principles schemes. Ballistic equilibrium conduction spectra and current-voltage characteristics prove that quantum conductance currents of these nanoribbon double heterostructures originate from the electron resonant tunneling between the topological edge-states of the two constituent Sb or Bi monolayer nanoribbons through the central barrier of $SbXH_n$ or $BiXH_n$ nanoribbon segment. This renders a high resonant current peak with strong negative differential conductance, thus being competent for zero-loss and ultrahigh-frequency resonant tunneling nanotransistors.

Keywords: first-principles calculation; two-dimensional topological material; quantum ballistic conductance; resonant tunneling transistor

1. Introduction

Nanostructures of two-dimensional (2D) materials are more sensitive to accurate processing due to their controllable electronic edge-states at atomic scale, implying their higher compactness and stability than three-dimensional (3D) materials in electronic device integration [1]. When silicon technology reaches the development of super-devices, the miniaturization of energy-saving transistors will focus on how to avoid the size limitation that leads to increased power consumption and decreased performance [2], in which 2D materials are expected to replace silicon-based 3D materials as electron transport channels [3]. However, since graphene-like materials even with an ultra-high electron mobility still lack a sufficient bulk electronic band-gap, the research focus of 2D topological materials has shifted to other 2D systems, such as indium selenide (InSe) and vanadium diselenide monolayer transistors [4,5]. Hexagonal graphene/boron-nitride sandwiched structure can realize electron resonant transitions by aligning Fermi levels of graphene electrodes, thus producing strong negative differential conductance with an extremely high peak-to-valley ratio, which is expected to be applied in ultrahigh-frequency and energy-saving electron transport devices [6]. In the 2D family of III-VI compounds, it has been demonstrated by both experimental material syntheses and electron-transport simulations that the carrier mobility of multi-layered InSe nanotransistor reaches ~2×10^3 $cm^2/(V{\cdot}s)$ at room temperature due to its high robustness to intra-band tunneling current, resulting in a higher open-to-switch current ratio than that of transition-metal dichalcogenides [7,8].

When the size of microelectronic components is reduced to less than 14 nm, the limitations to high-density microcircuit manufacturing technology and device performance can be overcome by developing new transistor channels [9,10]. The size of electron transport channel determines the global device size of electron transistor, which is inversely proportional to operation speed and energy efficiency [11,12]. The resonant tunneling transistor (RTT) made of 2D topological insulators (TIs) has attracted significant attention in recent years due to its great potential in applications of logic gates and signal processing devices [13–16]. Non-equilibrium Green's function combined with conjugate orbital tight-binding approach has been employed to calculate electron transports in graphene nanoribbon RRTs, which elucidated the effects of nanoribbon shape and operation temperature on electron resonant tunneling, and proved the versatile tuning functions realized by adjusting width and length of tunneling channel [13]. The graphene/boron-nitride(BN)/graphene nanoribbon double heterostructures represent an obvious resonant peak and the strong negative differential conductance in peak-valley ratios of 1~4, in which tunneling oscillation currents in radio frequencies can also be introduced for ultrahigh frequency devices [12,14]. Electron transports in the transversely repeated graphene/BN multi-heterostructures represent distinct characteristics of electron resonant tunneling [15]. When employing the graphdiyne nanoribbon with an extremely high electron mobility as electron transport channel in nanoribbon RTTs of graphene/graphdiyne/graphene double-heterostructures, the resonant current peak-valley ratio approaches to 4.5~6.0 [16].

Hexagonal honeycomb 2D elemental materials composed of antimony (Sb) or bismuth (Bi) element have a strong spin-orbit coupling and are expected to become a new family of 2D TIs at room temperature [17,18]. The Sb (111) atomic bi-layer can be converted into a two-dimensional topological phase by applying bi-axial strains in layer-plane, while Bi (111) bi-layer is a characteristic 2D topological insulator with a bulk band-gap of ~0.2 eV [19]. Although Bi monolayer 2D material has been experimentally prepared, its quantum spin hall effect has not been observed [20,21]. The 2D topological materials with large bulk band-gaps can be obtained through the surface chemical modification of halogenation or hydrogenation, but the plasma method used for material preparation leads to a hike in lattice defects, and even causes obvious impacts on topological properties [22]. In contrast, the surface chemical decorations by functional groups (surface functionalization) undergo a significantly slower reaction dynamic process, which is preferable to passivate surface conjugate bonds of 2D materials and increase bulk band-gap [23]. Furthermore, the hydrogenated surface is vulnerable to be oxidized at room temperature and ambient pressure, and has a poor chemical stability, while the chemical group functionalized surface presents an antioxidant ability and a higher thermal stability [24–26]. In addition to improving chemical stability and obtaining a sufficient bulk band-gap that can resist thermal excitation of electrons at room temperature, the surface chemical modifications can also ameliorate band structure and topological state of 2D topological materials [27–31].

In the present study, the nanoribbon electronic structures of methyl, amino, or hydroxyl surface-functionalized Sb and Bi monolayers ($SbXH_n$ and $BiXH_n$ monolayers: $XH_n = CH_3, NH_2, OH$) are investigated by first-principles calculations. By local surface functionalization on the central region of Sb or Bi monolayer nanoribbon by methyl, amino or hydroxyl group, the bipolar transport device models of $Sb/SbXH_n/Sb$ and $Bi/BiXH_n/Bi$ double-heterostructures are built up, for which electron transport characteristics (quantum ballistic conduction spectra and current-voltage curves) are calculated by nonequilibrium Green's function combined with first-principles method. It is objective of applying surface-functionalized Sb and Bi momolayer nanoribbons to RTT channels in the merit of designing and manufacturing ultra-fast nanotransistors with 2D topological materials.

2. Modelling and Calculation Methodology

Based on 2D periodic structures of Sb or Bi and $SbXH_n$ or $BiXH_n$ (surface modified by functional groups: $XH_n = CH_3, NH_2, OH$) monolayers, the nanoribbons with the two edges in zigzag atomic alignment (zigzag nanoribbons) are modeled for various widths, whose electronic band structures in one-dimensional Brillouin zone are calculated by first-principles all-electron numerical basis-set schemes, as implemented in Dmol3 code of Materials Studio package (Accelrys Inc., Materials Studio version 2020.08, San Diego, CA, USA). One-dimensional nanoribbons are modeled in virtual 3D periodic lattice: a ~30 Å vacuum layer is set between the two adjacent periodic images to completely avert electron wave-function overlaps between adjacent nanoribbons in 3D periodic lattice, which actually represents a nanoribbon one-dimensional structure. The constructed models of Sb or Bi and $SbXH_n$ or $BiXH_n$ monolayer zigzag nanoribbons are flanked by identical atom aligning sequences, thus possessing the similar space inversion symmetry as their host 2D monolayers. Nanoribbon width is evaluated by the number of hexagonal rings N_w of Sb or Bi atoms in nanoribbon lateral direction, namely the number of double Sb or Bi atoms in nanoribbon periodic unit. The Sb or Bi atoms at the two nanoribbon edges are chemically covered with hydrogen atoms to passivate the unsaturated edge bonds. Bipolar transport device models of $Sb/SbXH_n/Sb$ and $Bi/BiXH_n/Bi$ nanoribbon double-heterostructures are established to simulate RRTs, as shown in Figure 1: total length of central scattering region is persisted to 12 nanoribbon periodic units (electron scatterings at two electrodes cannot affect each other); the length of $SbXH_n$ or $BiXH_n$ nanoribbon segment is specified by N_L (the number of nanoribbon periodic units); nanoribbon orientation is specified along [100] crystallographic direction of host 2D monolayer (zigzag edge).

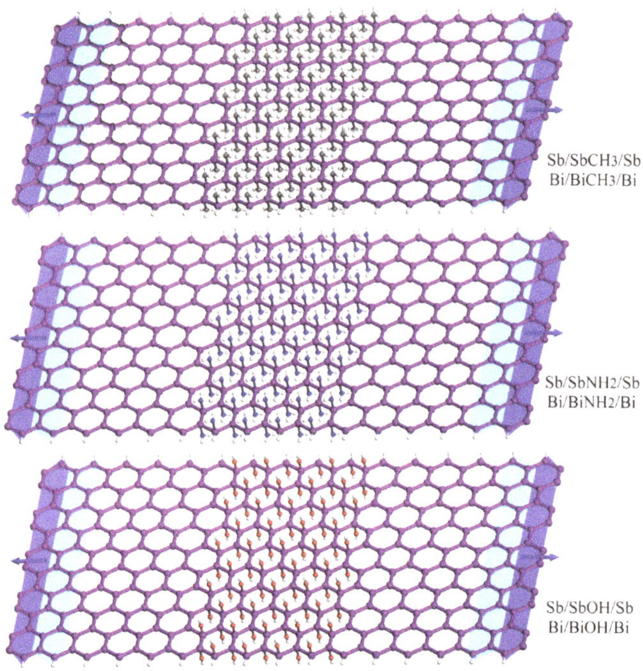

Figure 1. Bipolar transport device models of $Sb/SbXH_n/Sb$ or $Bi/BiXH_n/Bi$ double-heterostructure RRTs, where two semi-infinite electrodes are specified at two terminals of Sb or Bi monolayer nanoribbons. Purple, gray, blue, red and white spheres represent the chemically bonding atoms of Sb or Bi, carbon, nitrogen, oxygen and hydrogen respectively. Shallow blue and deep blue colors indicate buffer and electrode regions respectively.

Geometry optimizations of these nanoribbons and their double-heterostructures are performed by *Smart* iteration algorithm until iteration energy difference, atomic force and atom displacement are less than 0.02 kcal/mol, 0.1 kcal/mol/Å and 0.001 Å, respectively. Quantum ballistic conduction spectra (electron transmission spectra) and current versus voltage variation profiles of $Sb/SbXH_n/Sb$ and $Bi/BiXH_n/Bi$ nanoribbon double-heterostructures are calculated by non-equilibrium Green's function combined with first-principles all electron orbital basis method [32–34]. The specified schemes and parameter settings of Dmol3 code adopted to calculate band structure and electron transport are listed in Table 1.

Table 1. Specifications in band structure and electron transport calculations with Dmol3 code.

Electron-State Solution	Schemes	Specification
Exchange-Correlation Functional	Gradient Corrected Approximation	PBEsol [35]
Potential	Core Treatment	all electron relativistic [36]
Basis Set	Numerical Atom-Orbit	Double Polarization
Orbital Cutoff	Global	5.0 Å
Self-Consistent Field Iteration	Tolerance	1×10^{-6} Ha/atom (1 Ha = 27.2 eV)
	Multipolar Expansion	Hexadecapole
	Density Mixing	Charge Amplitude = 0.2
BZ Integration *k*-sampling (band structure)	Monkhorst-Pack Grid [37]	$1 \times 1 \times 10$
BZ Integration *k*-sampling (electron transport)	Uniform Grid	0.02 $Å^{-1}$
Poisson Solver [38] (electron transport)	Buffer Length	7.5 Å
	Poisson Grid Maximum Element	0.5 Å
	Electrode Boundary Condition	Dirichlet
	Side Boundary Condition	Neumann
	Electrode Buffer Length	3 Å

3. Results and Discussion

3.1. Electronic Band Structure

For Sb and Bi monolayer nanoribbons, the p_z-orbital of Sb or Bi atoms is laterally coupled into a huge delocalized conjugate π bond to form one-dimensional electron gas in monolayer plane. The conjugate π bonding electrons near the two edges of Sb or Bi monolayer nanoribbon give rise to a highly conductive channel along nanoribbon extension direction, which appears in band structure by multiple overlapping metallic bands crossing with Fermi level along **G-X** path of one-dimensional Brillouin zone, as shown in Figure 2. Although the nontrivial topology of Sb or Bi monolayer keeps their topological nanoribbon edge-states being protected for back-scattering forbidden, the Dirac-point of topological edge-states at Brillouin zone boundary **X** is significantly higher than Fermi level and cannot be incorporated into metallic bands of one-dimensional electron gas near Fermi level. Nevertheless, when nanoribbon width is raised to $N_w > 2$, the electronic-state density of Sb or Bi p_z-orbitals near Fermi level increases enough to cause the band of topological edge-states just below Fermi level elevating to Fermi level. Consequently, a Dirac cone is formed at Fermi level with a smaller effective electron mass than both the conjugate π bonds at Fermi level and the topological edge-states considerably higher than Fermi level, implying that a significant part of electron-states at Fermi level are under topological protection from back scattering, which is actually a nontrivial topology defined by Chern's number that will be spin-splitting in broken time-reversal symmetry under magnetic field as 2D free electron gas [39]. For an even wider Sb or Bi monolayer nanoribbon, this character is more evident in band structure.

As shown in Figure 3 for the calculated band structures of $SbXH_n$ or $BiXH_n$ monolayer zigzag nanoribbons, the Fermi level, especially for $BiCH_3$ nanoribbon, is overall ~1 eV higher than that of Sb or Bi monolayer zigzag nanoribbons. Surface functionalization on Sb and Bi monolayer nanoribbons causes a complete transformation of band structures near Fermi level, which manifests Z_2 invariant (time-reversal symmetry) topological properties derived from hexagonal honeycomb arrangement of Sb or Bi atoms. When $N_w > 4$, the Z_2 invariant topological edge-states on two nanoribbon edges cannot couple each other, and a separate Dirac-cone-like band dispersion is formed at Brillouin zone boundary **X** point near Fermi level which is slightly higher than Dirac-point. The coupling of topological edge-states between the two nanoribbon edges can open a electronic band-gap and decrease Fermi level, as indicated by the slightly increasing Fermi level with N_W at $N_W < 5$ and the constant Fermi level at $N_W > 4$.

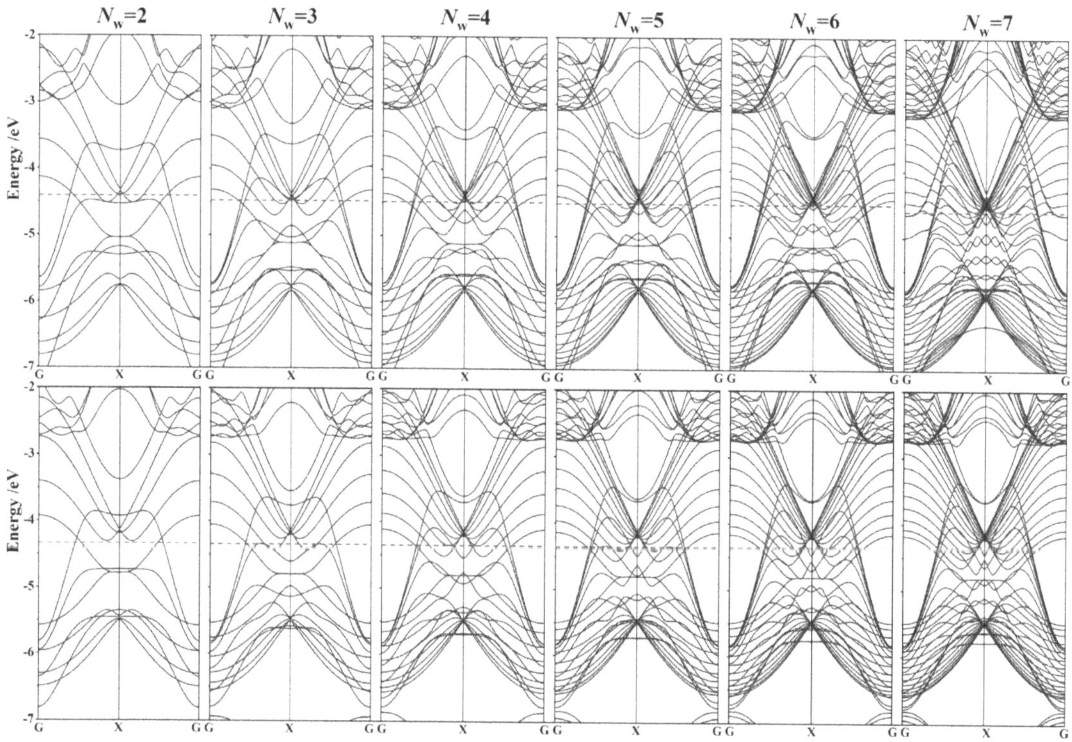

Figure 2. Band structures of Sb (**top panels**) and Bi (**bottom panels**) monolayer zigzag nanoribbons for diverse widths (N_W = 2~7). Vacuum free electron energy is set as energy zero and Fermi level is indicated by horizontal dash line.

The Fermi level of Sb or Bi monolayer nanoribbon decreases with the increase of N_W, as shown by the left panel in Figure 4. This proves that Fermi-level electron-states are partly derived from the coupled p_z-orbitals of the huge delocalized π bond which is subject to the quantum confinement by nanoribbon width, because the energy level of quantum-confined electrons should decrease with confinement dimension. On the contrary, the Fermi level of $SbXH_n$ or $BiXH_n$ monolayer nanoribbons increases slightly or persists unchanged with the increase of N_W, as shown by the right panel in Figure 4, indicating that Dirac point near Fermi level comes completely from Z_2 invariant topological edge-states which is distinguished from electron-states internal nanoribbons.

3.2. Electron Transport in Quantum Resonance

Given that the present nanoribbon double-heterostructure RRTs have mirror or inversion symmetry in atomic structures, it is equivalent of applying positive and negative voltages in opposite directions of bipolar device. Hence, one electrode is arbitrarily chosen as positive electrode and the other one as negative electrode to be applied with $U_b/2$ and $-U_b/2$ electrostatic potential, respectively, for the modeled double-heterostructure bipolar devices, where U_b is defined as bias voltage (or voltage bias) and specified in range of 0~1 V to calculate quantum ballistic conduction spectra (electron transmission spectra) and current versus bias voltage (I-U_b) profiles, as shown in Figure 5. A threshold voltage is defined here specifically as the bias voltage where the initial electric current rises to 1/10 of peak current. There is an inevitable Ohmic-like electric conduction due to the coupling between high-mobility topological edge-states of Sb or Bi and $SbXH_n$ or $BiXH_n$ nanoribbons at $Sb/SbXH_n$ or $Bi/BiXH_n$ interfaces (heterojunctions), which can be intensified by increasing electric field at these heterojunctions but will approach an uppermost limit at a bias voltage below resonant tunneling threshold.

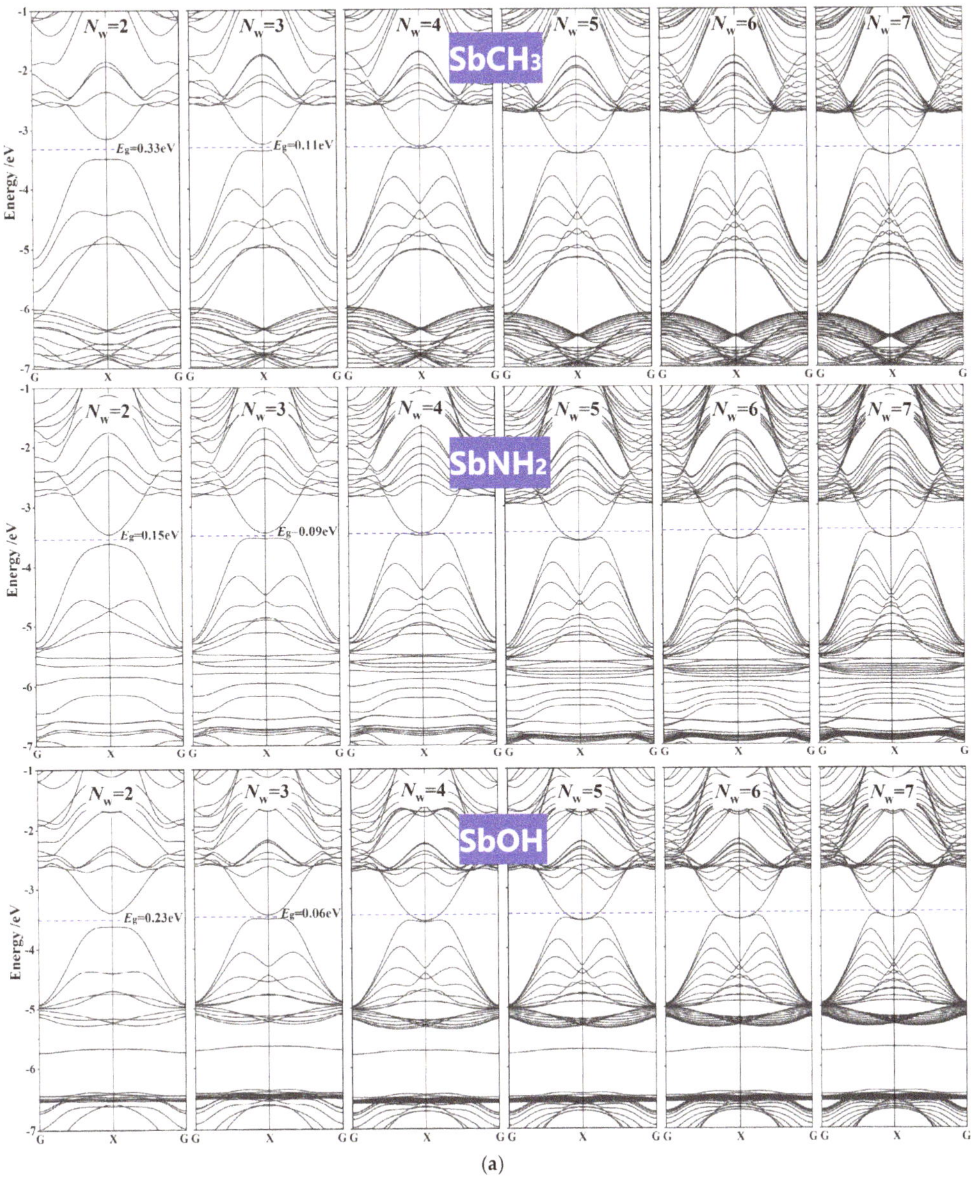

(a)

Figure 3. *Cont.*

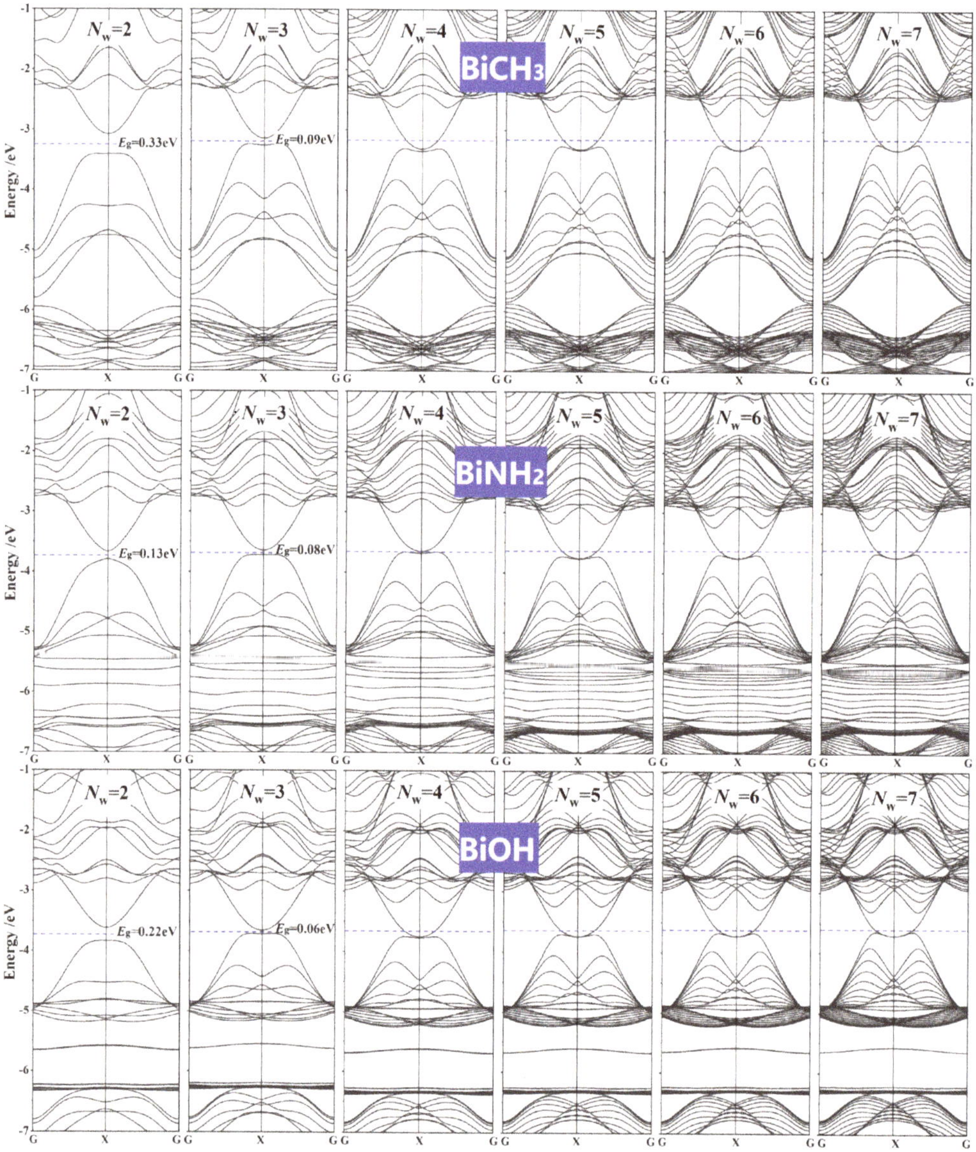

(b)

Figure 3. Band structures of (**a**) $SbXH_n$ and (**b**) $BiXH_n$ monolayer zigzag nanoribbons (N_w = 2~7) with their surfaces being chemically decorated by methyl, amino or hydroxyl group. Vacuum free electron energy is set as energy zero and Fermi level is indicated by horizontal dash line, and E_g denotes energy band gap.

When U_b is lower than threshold voltage (U_{th} = 0.3~0.4 V), the quantum conductance current is still smaller than 1 μA. With the increase of U_b, the Fermi level contributed by topological edge-states of Sb or Bi monolayer nanoribbon segment in the relatively low electric potential area (source region) elevates gradually in these double-heterostructure RRTs, and the triangular energy barrier formed by the quasi Fermi level in central $SbXH_n$

or $BiXH_n$ nanoribbon segment (tunneling channel) becomes more accurate, which means a decreasing length of tunneling channel. Accordingly, when U_b exceeds U_{th}, electron resonant tunneling occurs between the topological edge-states of the two constituent Sb or Bi monolayer nanoribbons, and the quantum conductance current increases exponentially with the increase of U_b. Since the transporting electrons near Fermi level are dominantly derived from nanoribbon topological edge-states, for which the back-scattering is forbidden by topological protection, the mobility of electron transport is completely determined by effective electron mass of topological edge-states in these double-heterostructure RRTs. Because the Dirac point of Sb or Bi monolayer nanoribbons becomes more accurate after surface chemical modification by methyl, amino or hydroxyl, that is, the effective electron mass near Fermi level is substantially reduced, the electron mobility of Sb or Bi monolayer nanoribbons is lower than that $SbXH_n$ or $BiXH_n$ monolayer nanoribbons, which determines the quantum conductance currents of $Sb/SbXH_n/Sb$ or $Bi/BiXH_n/Bi$ double-heterostructure RRTs. Without gate voltage, the electron resonant tunneling in these nanoribbon double-heterostructures produces the resonant current peak at U_b = 0.55~0.8 V; as bias voltage further increases, the topological conductive channels of the two constituent Sb or Bi nanoribbons in these double-heterostructures begin to stagger, which accounts for the sharply decreasing electron transmission probability and resonant tunneling current with bias voltage, resulting in an extremely high negative differential conductivity.

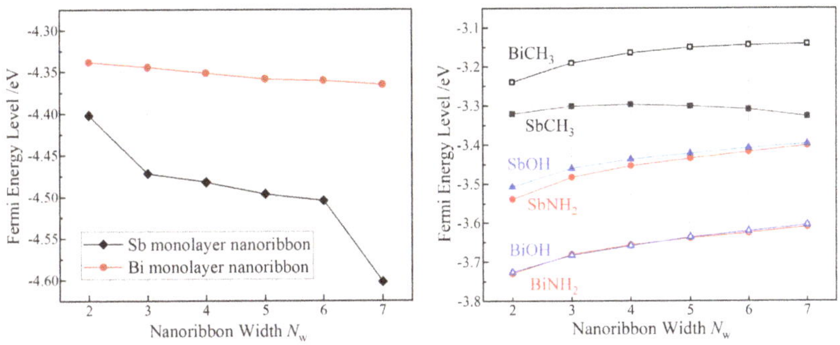

Figure 4. Fermi energy levels in dependence on nanoribbon width (N_w = 2~7) of Sb or Bi (**left panel**) and $SbXH_n$ or $BiXH_n$ (**right panel**) monolayer zigzag nanoribbons in reference to vacuum free electron energy level.

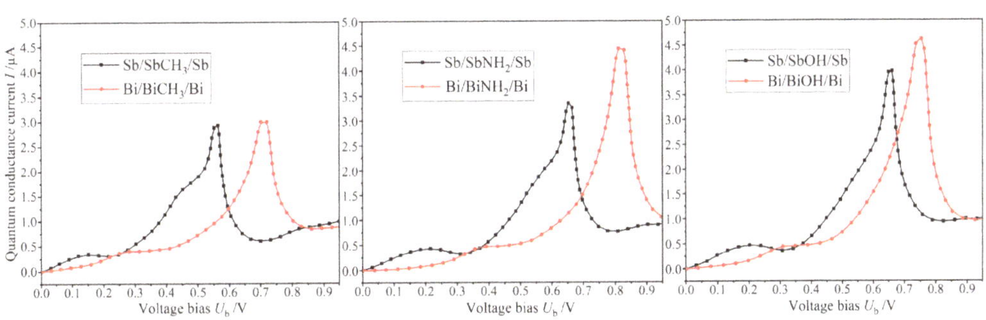

Figure 5. Quantum conductance I-U_b characteristics of resonant tunneling transistors modeled by $Sb/SbXH_n/Sb$ and $Bi/BiXH_n/Bi$ double-heterostructures of monolayer zigzag nanoribbons.

As shown by the 0.0~1.0 V electron transmission spectra of $Sb/SbXH_n/Sb$ and $Bi/BiXH_n/Bi$ nanoribbon double-heterostructures in Figure 6, electron resonant tunneling arises when bias voltage is raised to 0.4 V, producing electron transmission peaks in $(-U_b/2, +U_b/2)$ bias window, accounting for the quantum conductance current increasing exponentially with voltage bias. As bias voltage increases to 0.55~0.65 V, the Fermi levels of topological electrons at the two electrodes of Sb/SbXH_n/Sb nanoribbon double-heterostructure bipolar devices (RRTs) begin to separate each other when the resonant tunneling turns to ordinary electron tunneling, as manifested by the electron transmission peak near Fermi level shifting out of bias window, which results in a prompt abatement in quantum conductance current and a strong negative differential conductance (NDC) with resonant peak-valley ratios (PVR) of 4.3~4.9. A higher PVR means a longer lifetime of transporting electrons, according to which the optimal working frequency can be effectively set for RRTs. In comparison, Bi/BiXH_n/Bi nanoribbon double-heterostructures give rise to strong NDC with PVR of 3.2~4.8 under voltage bias of 0.7~0.8 V.

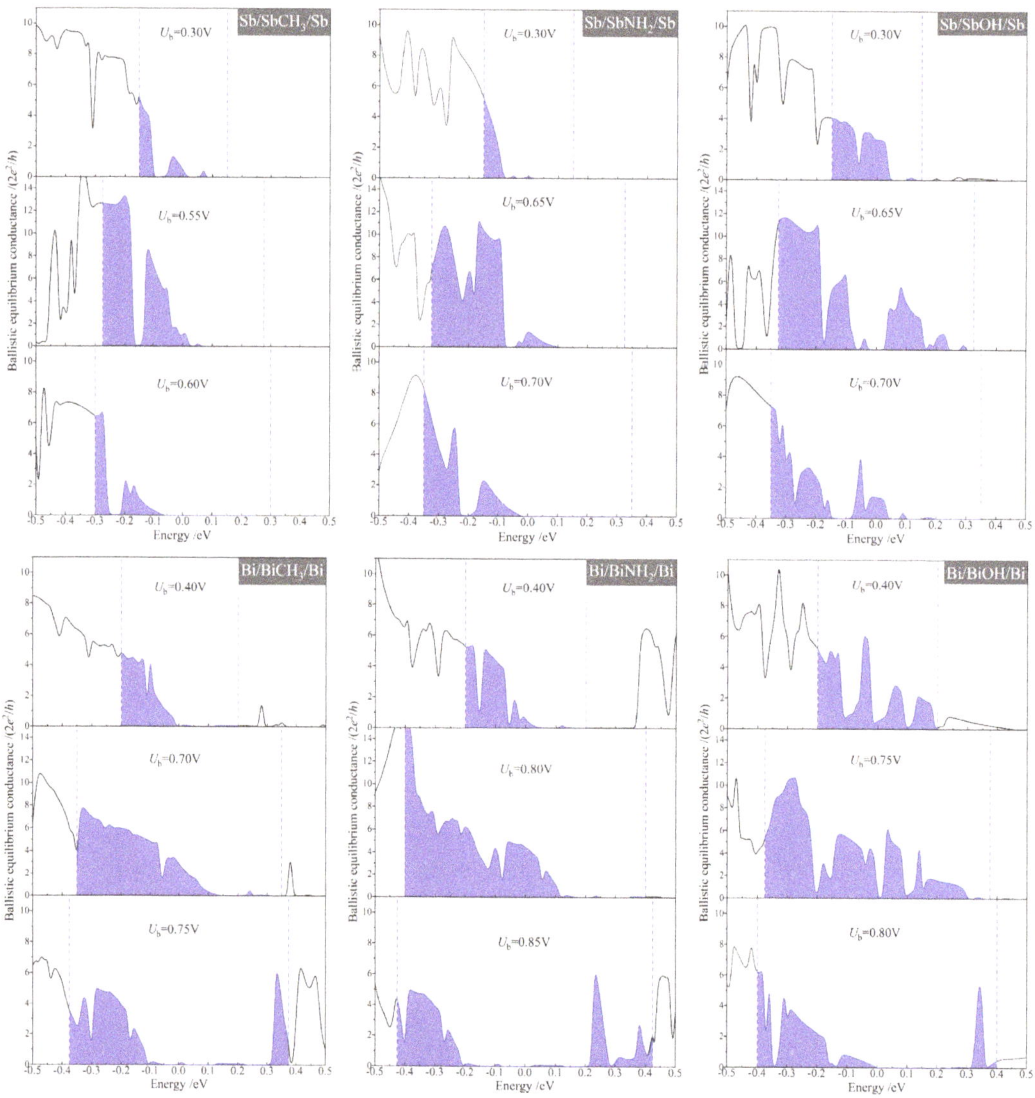

Figure 6. Electron transmission spectra of resonant tunneling transistors modeled by Sb/SbXH_n/Sb and Bi/BiXH_n/Bi nanoribbon double-heterostructures.

It has been verified from a recent report that the resonant tunneling current increases sensitively with channel length, whilst persisting on a constant PVR of NDC in BP/silicon-carbide(SiC) nanoribbon double-heterostructure RRT [40]. As illustrated in Figures 2 and 3, in contrast to Sb or Bi monolayer nanoribbons, the $SbXH_n$ or $BiXH_n$ monolayer nanoribbons exhibit a more accurate Dirac-point of topological edge-states with a lower effective electron mass. Accordingly, it is reasonably suggested for the electron transport characteristics of Sb/$SbXH_n$/Sb and Bi/$BiXH_n$/Bi nanoribbon double-heterostructures as follows: the highly conductive tunneling channel through $SbXH_n$ or $BiXH_n$ barrier is dominantly derived from nanoribbon topological edge-states, not from quantum confined electrons internal nanoribbons, which are under topological protection from back-scattering in each constituent nanoribbon segment of double-heterostructures; given $N_W > 4$, the coupling between topological electron-states of the two nanoribbon edges can be avoided, so that resonant tunneling current is independent of nanoribbon width; electron scattering primarily occurs at Sb/$SbXH_n$ or Bi/$BiXH_n$ heterojunction interfaces, which is determined by energy barrier height and length of central $SbXH_n$ or $BiXH_n$ monolayer nanoribbon segments.

Therefore, the resonant tunneling current peaks of Sb/$SbXH_n$/Sb and Bi/$BiXH_n$/Bi nanoribbon double-heterostructures decrease exponentially and linearly, respectively, with the increase of N_L, while PVR of NDC relies merely on the species of functional group decorated on Sb or Bi nanoribbon surfaces (the higher oxide-ability of functional group used for surface chemical decoration results in a higher electron barrier and PVR), as shown in Figure 7. The recently reported PVR of the nanoribbon RRTs constituted with different monolayer materials are listed in Table 2, indicating that Sb/$SbXH_n$/Sb and Bi/$BiXH_n$/Bi nanoribbon double-heterostructures have higher PVR values than graphene/BN/graphene and GaN/Al/GaN nanoribbon double-heterostructures, which implies they are more efficient nanochannels for RTTs.

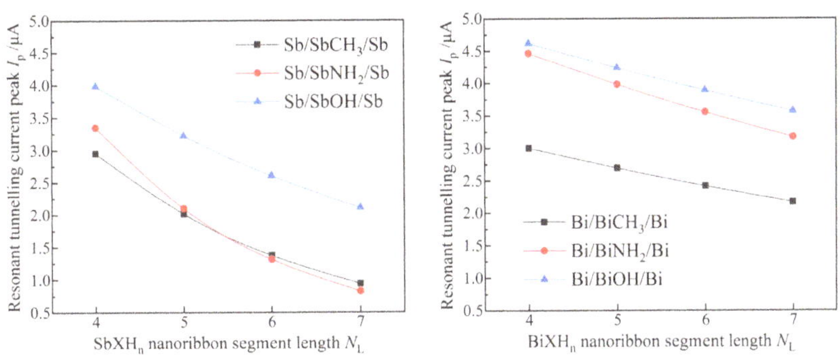

Figure 7. Quantum conductance current peaks (I_p) of electron resonant tunneling for Sb/$SbXH_n$/Sb (**left panel**) and Bi/$BiXH_n$/Bi (**right panel**) nanoribbon double-heterostructures (N_W = 5) in dependence on the length (N_L) of central $SbXH_n$ or $BiXH_n$ nanoribbon segments.

Table 2. Threshold gate voltage and PVR for resonant tunneling transistors of nanoribbon double-heterostructures constructed by specific 2D monolayer materials.

Nanoribbon Scattering Region	Nanoribbon Electrodes	Research Approach	Gate Voltage/V	PVR	Data Source
Sb/$SbXH_n$/Sb	Sb	First-principles calculations	0	4.3~4.9	Present study
Bi/$BiXH_n$/Bi	Bi	First-principles calculations	0	3.2~4.8	Present study
BN	Graphene	Theoretical calculations and experiments	0, 20	1~4	Ref. [14]
BN	Graphene	Theoretical calculations and experiments	−0, ±40	–	Ref. [12]
Graphdiyne	Graphene	First-principles calculations	5	4.5, 6.0	Ref. [16]

4. Conclusions

Fermi levels of Sb or Bi monolayer nanoribbons are determined by the infinite delocalized π bond similar to one-dimensional electron gas, while nanoribbon topological edge-states arise at Brillouin zone boundary **X** point with evidently higher or slightly lower energy levels than Fermi level. Topological edge-states of Sb or Bi monolayer nanoribbon can be merged into Fermi level by increasing nanoribbon width and thus contribute to electron transports with an extremely high mobility under partly topological protection. Surface functionalized Sb and Bi monolayer nanoribbons present a pure Dirac-cone near Fermi level derived from nanoribbon topological edge-states, which accounts for electron transports completely under topological protection without back-scattering. $Sb/SbXH_n/Sb$ and $Bi/BiXH_n/Bi$ nanoribbon double-heterostructures render a significant resonant current peak with a strong negative differential conductance (NDC) in a narrow voltage bias range by the resonant tunneling of nanoribbon topological edge-states. Resonant current peaks of these simulated double-heterostructure resonant tunneling transistors increase exponentially and linearly with the lengths of central $SbXH_n$ and $BiXH_n$ tunneling channels, respectively. In particular, the peak-valley ratio of NDC only depends on the species of functional group used for surface chemical decoration on central tunneling channel and is not affected by channel length. Quantum conductance currents of $Sb/SbXH_n/Sb$ and $Bi/BiXH_n/Bi$ nanoribbon double-heterostructures are generated from the resonant tunneling of high mobility topological edge-states between the two constituent Sb or Bi nanoribbons, which are independent of nanoribbon width and give maximum NDC peak-valley ratios of 4.9 and 4.8 respectively. The present study suggests that surface-functionalized Sb and Bi monolayers are potential in future applications of ultrahigh-frequency and zero-loss topological nanotransistors.

Author Contributions: Data curation, formal analysis, writing—original draft preparation, writing—review and editing, Q.G.; conceptualization, investigation, project administration, G.Z. All authors have read and agreed to the published version of the manuscript.

Funding: This research is funded by the National Natural Science Foundation of China (Grant No. 51973046).

Institutional Review Board Statement: Not applicable.

Informed Consent Statement: Not applicable.

Data Availability Statement: Theoretical methods and results are available from all the authors.

Conflicts of Interest: The authors declare no conflict of interest.

References

1. Zeng, H.; Zhi, C.; Zhang, Z.; Wei, X.; Wang, X.; Guo, W.; Bando, Y.; Golberg, D. "White Graphenes": Boron Nitride Nanoribbons via Boron Nitride Nanotube Unwrapping. *Nano Lett.* **2010**, *10*, 5049–5055. [CrossRef] [PubMed]
2. Ieong, M.; Doris, B.; Kedzierski, J.; Min, Y. Silicon Device Scaling to the Sub-10-nm Regime. *Science* **2004**, *306*, 2057–2060. [CrossRef] [PubMed]
3. Scalise, E.; Iordanidou, K.; Afanas'ev, V.V.; Stesmans, A.; Houssa, M. Silicene on non-metallic substrates: Recent theoretical and experimental advances. *Nano Res.* **2018**, *11*, 1169–1182. [CrossRef]
4. Ho, P.H.; Chang, Y.R.; Chu, Y.C.; Li, M.K.; Tsai, C.A.; Wang, W.H.; Ho, C.H.; Chen, C.W.; Chiu, P.W. High-Mobility InSe Transistors: The Role of Surface Oxides. *ACS Nano* **2017**, *11*, 7362–7370. [CrossRef] [PubMed]
5. Lv, M.H.; Li, C.M.; Sun, W.F. Spin-Orbit Coupling and Spin-Polarized Electronic Structures of Janus Vanadium-Dichalcogenide Monolayers: First-Principles Calculations. *Nanomaterials* **2022**, *12*, 382. [CrossRef] [PubMed]
6. Mishchenko, A.; Tu, J.S.; Cao, Y.; Gorbachev, R.V.; Wallbank, J.R.; Greenaway, M.T.; Morozov, V.E.; Morozov, S.V.; Zhu, M.J.; Wong, S.L.; et al. Twist-controlled resonant tunnelling in graphene/boron nitride/graphene heterostructures. *Nat. Nanotechnol.* **2014**, *9*, 808–813. [CrossRef] [PubMed]
7. Sucharitakul, S.; Goble, N.J.; Kumar, U.R.; Sankar, R.; Bogorad, Z.A.; Chou, F.C.; Chen, Y.T.; Gao, X.P.A. Intrinsic Electron Mobility Exceeding 10^3 $cm^2/(V{\cdot}s)$ in Multilayer InSe FETs. *Nano Lett.* **2015**, *15*, 3815–3819. [CrossRef]
8. Marin, E.G.; Marian, D.; Iannaccone, G.; Fiori, G. First-Principles Simulations of FETs Based on Two-Dimensional InSe. *IEEE Electron. Device Lett.* **2018**, *39*, 626–629. [CrossRef]

9. Pi, S.; Lin, P.; Xia, Q. Fabrication of sub-10 nm metal nanowire arrays with sub-1nm critical dimension control. *Nanotechnology* **2016**, *27*, 464004–464009. [CrossRef]
10. Lawrence, T.C.; Vashishtha, V.; Shifren, L.; Gujja, A.; Sinha, S.; Cline, B.; Ramamurthy, C.; Yeric, G. ASAP7: A 7-nm film FET predictive process design kit. *Microelect. J.* **2016**, *53*, 105–115.
11. Guo, Y.G.; Wang, F.Q.; Wang, Q. An all-carbon vdW heterojunction composed of penta-graphene and graphene: Tuning the Schottky barrier by electrostatic gating or nitrogen doping. *Appl. Phys. Lett.* **2017**, *111*, 073503. [CrossRef]
12. Punniyakoti, S.; Sivakumarasamy, R.; Vaurette, F.; Joseph, P.; Nishiguchi, K.; Fujiwara, A.; Clement, N. Hydrogen Silsesquioxane-Based Nanofluidics. *Adv. Mater. Interface.* **2017**, *4*, 1601155. [CrossRef]
13. Teong, H.; Lam, K.T.; Khalid, S.B.; Liang, G. Shape effects in graphene nanoribbon resonant tunneling diodes: A computational study. *J. App. Phys.* **2009**, *105*, 084317–084321. [CrossRef]
14. Britnell, L.; Gorbachev, R.V.; Geim, A.K.; Ponomarenko, L.A.; Mishchenko, A.; Greenaway, M.T.; Fromhold, T.M.; Novoselov, K.S.; Eaves, L. Resonant tunnelling and negative differential conductance in graphene transistors. *Nat. Commun.* **2013**, *4*, 1794. [CrossRef] [PubMed]
15. Ongun Özçelik, V.; Durgun, E.; Ciraci, S. Modulation of electronic properties in laterally and commensurately repeating graphene and boron nitride composite nanostructures. *J. Phys. Chem. C* **2015**, *119*, 13248–13256. [CrossRef]
16. Zhang, S.T.; Sun, Z.; Zhao, L. First-principles study of graphene nanoflakes with large spin property. *Act. Phys. Sin.* **2018**, *67*, 187102. [CrossRef]
17. Zhang, P.F.; Liu, Z.; Duan, W.; Liu, F.; Wu, J. Topological and electronic transitions in a Sb(111) nanofilm: The interplay between quantum confinement and surface effect. *Phys. Rev. B* **2013**, *85*, 201410. [CrossRef]
18. Huang, Z.Q.; Chuang, F.C.; Hsu, C.H.; Liu, Y.T.; Chang, H.R.; Lin, H.; Bansil, A. Nontrivial topological electronic structures in a single Bi(111) bilayer on different substrates: A first-principles study. *Phys. Rev. B* **2013**, *88*, 165301. [CrossRef]
19. Wada, M.; Murakami, S.; Freimuth, F.; Bihlmayer, G. Localized edge states in two-dimensional topological insulators: Ultrathin Bi films. *Phys. Rev. B* **2011**, *83*, 121310. [CrossRef]
20. Yang, Z.Q.; Jia, J.F.; Qian, D. Two-dimensional topological insulators with large bulk energy gap. *Chin. Phys. B* **2016**, *25*, 117312. [CrossRef]
21. Wang, Z.F.; Yao, M.Y.; Ming, W.; Miao, L.; Zhu, F.; Liu, C.; Gao, C.L.; Qian, D.; Jia, J.F.; Liu, F. Creation of helical Dirac fermions by interfacing two gapped systems of ordinary fermions. *Nat. Commun.* **2013**, *4*, 1384. [CrossRef] [PubMed]
22. Wu, J.; Xie, L.M.; Li, Y.; Wang, H.; Ouyang, Y.; Guo, J.; Dai, H.G. Controlled chlorine plasma reaction for noninvasive graphene doping. *J. Am. Chem. Soc.* **2011**, *133*, 19668–19671. [CrossRef] [PubMed]
23. Jiang, S.; Krymowsk, K.; Asel, T.; Arguilla, M.Q.; Cultrara, N.D.; Yanchenko, E.; Yang, X.; Brillson, L.J.; Windl, W.; Goldberger, J.E. Tailoring the Electronic Structure of Covalently Functionalized Germanane via the Interplay of Ligand Strain and Electronegativity. *Chem. Mater.* **2016**, *28*, 8071–8077. [CrossRef]
24. Wu, Z.S.; Ren, W.; Gao, L.; Zhao, J.; Chen, Z.; Liu, B.; Tang, D.; Yu, B.; Jiang, C.; Cheng, H.M. Synthesis of Graphene Sheets with High Electrical Conductivity and Good Thermal Stability by Hydrogen Arc Discharge Exfoliation. *ACS Nano* **2009**, *3*, 411–417. [CrossRef] [PubMed]
25. Bagri, A.; Grantab, R.; Medhekar, N.V.; Sheno, V.B. Stability and Formation Mechanisms of Carbonyl- and Hydroxyl-Decorated Holes in Graphene Oxide. *J. Phys. Chem. C* **2010**, *114*, 12053–12061. [CrossRef]
26. Ma, Y.D.; Dai, Y.; Kou, L.Z.; Frauenheim, T.; Heine, T. Robust two-dimensional topological insulators in methylfunctionalized bismuth, antimony, and lead bilayer films. *Nano Lett.* **2015**, *15*, 1083–1089. [CrossRef]
27. Jiang, S.S.; Butler, S.; Bianco, E.; Restrepo, O.D.; Windl, W.; Goldberger, J.E. Improving the stability and optical properties of germanane via one-step covalent methyl-termination. *Nat. Commun.* **2014**, *5*, 3389. [CrossRef]
28. Qiu, J.; Fu, H.; Xu, Y.; Zhou, Q.; Meng, S.; Li, H.; Chen, L.; Wu, K. From Silicene to Half-Silicane by Hydrogenation. *ACS Nano* **2015**, *9*, 11192–11199. [CrossRef]
29. Özçelik, V.O.; Cahangirov, S.; Ciraci, S. Stable Single-Layer Honeycomblike Structure of Silica. *Phys. Rev. Lett.* **2014**, *112*, 246803. [CrossRef]
30. Knapp, D.; Brunschwig, B.S.; Lewis, N.S. Chemical, Electronic, and Electrical Properties of Alkylated Ge(111) Surfaces. *J. Phys. Chem. C* **2010**, *114*, 12300–12307. [CrossRef]
31. Gong, Q.; Zhang, G. Spin-Orbit Coupling Electronic Structures of Organic-Group Functionalized Sb and Bi Topological Monolayers. *Nanomaterials* **2022**, *12*, 2041. [CrossRef] [PubMed]
32. Madsen, G.K.H.; Singh, D.J. BoltzTraP. A code for calculating band-structure dependent quantities. *Comput. Phys. Commun.* **2006**, *175*, 67–71. [CrossRef]
33. Gunst, T.; Markussen, T.; Stokbro, K.; Brandbyge, M. First-principles method for electron-phonon coupling and electron mobility: Applications to two-dimensional materials. *Phys. Rev. B* **2016**, *93*, 035414. [CrossRef]
34. Pecchia, A.; Penazzi, G.; Salvucci, L.; Di Carlo, A. Non-equilibrium Green's functions in density functional tight binding: Method and applications. *New J. Phys.* **2008**, *10*, 065022. [CrossRef]
35. Perdew, J.P.; Ruzsinszky, A.; Csonka, G.I.; Vydrov, O.A.; Scuseria, G.E.; Constantin, L.A.; Zhou, X.; Burke, K. Restoring the Density-Gradient Expansion for Exchange in Solids and Surfaces. *Phys. Rev. Lett.* **2008**, *100*, 136406. [CrossRef]
36. Kadek, M.; Repisky, M.; Ruud, K. All-electron fully relativistic Kohn-Sham theory for solids based on the Dirac-Coulomb Hamiltonian and Gaussian-type functions. *Phys. Rev. B* **2019**, *99*, 205103. [CrossRef]

37. Monkhorst, H.J.; Pack, J.D. Special points for Brillouin-zone integrations. *Phys. Rev. B* **1976**, *13*, 5188. [CrossRef]
38. Ghanbari Shohany, B.; Roknabadi, M.R.; Kompany, A. Electrical investigation of armchair graphene-graphdiyne-graphene nanoribbons heterojunctions. *Commun. Theor. Phys.* **2016**, *65*, 99–104. [CrossRef]
39. Chang, C.Z.; Zhang, J.S.; Feng, X.; Shen, J.; Zhang, Z.C.; Guo, M.H.; Li, K.; Ou, Y.B.; Wei, P.; Wang, L.L.; et al. Experimental Observation of the Quantum Anomalous Hall Effect in a Magnetic Topological Insulator. *Science* **2013**, *340*, 167–170. [CrossRef]
40. Dong, J.; Li, H.; Li, L. Multi-functional nano-electronics constructed unsing boron phosphide and silicon carbide nanoribbons. *NPG Asia Mater.* **2013**, *5*, e56. [CrossRef]

Disclaimer/Publisher's Note: The statements, opinions and data contained in all publications are solely those of the individual author(s) and contributor(s) and not of MDPI and/or the editor(s). MDPI and/or the editor(s) disclaim responsibility for any injury to people or property resulting from any ideas, methods, instructions or products referred to in the content.

crystals

MDPI

Article

Modeling 2D Arrangements of Graphene Nanoribbons

Talia Tene [1], Marco Guevara [2], Nataly Bonilla García [3], Myrian Borja [3] and Cristian Vacacela Gomez [4,*]

1 Department of Chemistry, Universidad Técnica Particular de Loja, Loja 110160, Ecuador
2 Faculty of Mechanical Engineering, Escuela Superior Politécnica de Chimborazo (ESPOCH), Riobamba 060155, Ecuador
3 Escuela Superior Politécnica de Chimborazo (ESPOCH), Riobamba 060155, Ecuador
4 INFN-Laboratori Nazionali di Frascati, Via E. Fermi 54, I-00044 Frascati, Italy
* Correspondence: cristianisaac.vacacelagomez@fis.unical.it

Abstract: In the last two decades, interest in graphene has grown extensively due to its extraordinary properties and potential for various applications such as sensing and communication. However, graphene is intrinsically a semimetal with a zero bandgap, which considerably delays its use where a suitable bandgap is required. In this context, quasi-one-dimensional counterparts known as graphene nanoribbons (GNRs) have demonstrated sizeable bandgaps and versatile electronic properties, which make them promising candidates for photonic and plasmonic applications. While progress has recently been made toward the synthesis of GNRs, theoretical models to envisage their electronic and optical properties have been restricted to ab initio approaches, which are not feasible for wide systems because of the large number of atoms tangled. Here, we use a semi-analytical model based on Dirac cone approximation to show the adjustable electronic and plasmonic characteristics of wide and experimental GNRs, both freestanding and non-freestanding. This approach utilizes the group velocity of graphene, which is calculated using density functional computations ($v_F = 0.829 \times 10^6$ m s^{-1}), as the primary input. Importantly, our research reveals that at the terahertz level, the plasmon-momentum dispersion is highly responsive to changes by varying the ribbon width or charge carrier concentrations, the other involved parameters can be manipulated by setting values from experiments or more sophisticated predictions. In particular, this model can replicate the electronic properties of GNRs on Ge(001) and GNRs on Au(111). From the plasmonic side, the plasmon spectrum of graphene microribbon arrays of 4 µm wide on Si/SiO_2 and GNR arrays on Si are found in good agreement with experiments. The potential use of GNRs in sensing molecules such as chlorpyrifos-methyl is also discussed. Chlorpyrifos-methyl is chosen as the test molecule because it is a commonly used insecticide in agriculture, but its high toxicity to organisms and humans makes it a concern. It has been established that the plasmon resonances of all the studied GNRs occur at the same frequency as chlorpyrifos-methyl, which is 0.95 THz. Our findings can serve as a useful guide for future experiments.

Keywords: graphene nanoribbons; DFT; semi-analytical framework; THz

1. Introduction

Graphene is a planar sheet of sp^2-bonded carbon atoms tightly packed into a honeycomb-like lattice [1,2]. The carbon atoms form a two-dimensional (2D) hybridized network with three nearest neighbors, each at a distance of about 1.420 Å, leaving one unhybridized half-filled p-orbital [3], perpendicular to the graphene plane for every carbon atom. Since 2004 [4], graphene has been the subject of significant research and development, due to its extraordinary electronic, mechanical, thermal, and optical properties [5]. Indeed, graphene has shown a lot of promise for the replacement of silicon-based electronic devices [6]. However, as-found graphene is gapless and cannot be used immediately in up-to-date semiconductor technology because this carbon nanomaterial needs current on/off ratios. Consequently, a great deal of effort is the opening of a bandgap in graphene without significantly affecting its exceptional properties, particularly, its high electron mobility.

Within this frame of reference, several chemical methods have been proposed, for instance, the oxidation-reduction of graphite to produce graphene derivatives that show optical bandgaps larger than 2 eV [7] or graphenes from microwave-derived carbon precursors for energy savings of up to 60% [8]. These graphene derivatives have the potential for environmental applications [9], advanced composite, and smart (multifunctional) eco-composites [10–12] with applicability in future nanodevices [13,14]. Alternatively, a sizable bandgap can be opened up through the quantum confinement effect by patterning graphene in the form of graphene nanoribbons (GNRs) [15,16]. GNRs are thin and elongated strips with straight edges. As the ribbon width decreases, GNRs can gradually transform from semimetals to semiconductors [17]. As an example, a bandgap of about 100 meV is expected for a ribbon of 10 nm, whereas a ribbon of width ~2 nm would have a bandgap of about 1 eV. From a piece of important evidence, Fei Z et al., [18] have reported the successful preparation of GNR samples with widths of $w = 155$, 270, 380, and 480 nm. Most importantly, based on nano-infrared images, the authors showed the existence of two plasmon modes in experimental GNRs organized as 2D periodic arrays: the conventional surface plasmon and a novel confined (edge) plasmon. Both plasmon modes have been theoretically confirmed in ultra-narrow GNR systems working in the reciprocal space by an ab initio method [19–21].

Plasmons are oscillations of valence electrons on the surface of various materials such as metals, semiconductors, and topological insulators that occur collectively [22]. These collective oscillations can be utilized to control and focus optical energy on a nanometer scale. This is due to the strong interaction between these oscillations and light [23]. In particular, plasmons in graphene reveal high confinement, electrostatic tunability, and long lifetimes [24]. These plasmon features are promising for photonic applications from the terahertz- to the infrared regime. Additional control over surface plasmons has been observed in graphene nanoribbons (GNRs) because of (i) the change in dimensionality from 2D graphene to 1D graphene ribbons and (ii) the confinement of charge carriers [20]. This extra control is expected to be more significant than that of edge plasmons, which is dependent on the bandgap of the GNR [19]. These unique plasmonic properties of GNRs could be utilized to enhance current biosensors that utilize surface plasmon resonance and graphene [25–30].

Previous studies [31,32] have mainly used Time-Dependent Density Functional Theory (TDDFT) within the Random Phase Approximation (RPA) to investigate the plasmonic properties of ultra-narrow GNR arrays (<2 nm). However, this atomistic approach is not suitable for GNRs of the widths reported in Ref. [18]. In this work, we use an alternative method to examine the electronic and plasmonic properties of experimental GNRs. The proposed semi-analytical model [33] is based on the group velocity of graphene, which is calculated using density functional theory (DFT), as the main input. Both freestanding and non-freestanding systems are considered. We show that the plasmon frequency in wide GNRs is highly sensitive. Additionally, we explore the possibility of using it for sensing chlorpyrifos-methyl. The reason it was picked as the test molecule is that it is widely used in crops as an insecticide, but its harmful effects on living beings and humans raise concerns. Our results can guide the design of future nanodevices based on graphene.

2. Theoretical Framework

It is worth emphasizing that the current modeling approach focuses on the electronic and plasmonic properties of GNRs with widths that have been experimentally observed ($w = 155$, 270, 380, and 480 nm) [34,35]. The semi-analytical model is composed of two steps:

(i) Numerical computations using DFT to obtain an accurate group velocity of graphene.
(ii) Analytical expressions derived mainly from the Dirac cone approximation [33] to investigate the related electronic and plasmonic features.

2.1. DFT Computations

The ground-state properties of graphene were determined by plane-wave DFT computations within the Kohn–Sham formalism implemented in the Abinit software [36]. Electronic structure calculations were performed at the level of local density approximation (LDA) [37] using norm-conserving pseudopotentials to eliminate the core electrons [38]. The plane-wave basis is represented by the space functions:

$$PW_{\mathbf{k}+\mathbf{G}}(\mathbf{r}) = \Omega_0^{-1/2} e^{i(\mathbf{k}+\mathbf{G})\cdot\mathbf{r}} \tag{1}$$

where $\mathbf{k}$ is the wave vector in the first Brillouin zone (ΓKM path) (Figure S1a), $\mathbf{G}$ is the reciprocal-lattice vector, and Ω_0 is the unit-cell volume.

The number of plane waves was limited by the energy-cutoff $|\mathbf{k}+\mathbf{G}|/2 \leq 25$ hartrees (~680 eV). The three-dimensional periodicity required by plane-wave DFT was created by repeating the graphene sheets with a distance of 15 Å. Structural optimization was not taken into account in this case as the lattice constant was fixed at 2.460 Å (the case of ideal graphene). The ground-state calculations were carried out using a Γ—centered and unshifted Monkhorst–Pack grid [39] of 90 × 90 × 1 to obtain the complete band structure and density of states (DOS) of graphene (Figure S1b,c), including up to 20 bands. Additionally, a high-resolution Monkhorst–Pack grid of 720 × 720 × 1 was used to obtain the linear band structure of graphene around the K point (Figure S2a), including up to eight bands.

A close picture of the band structure (Figure S1b) shows that graphene exhibits the well-known Dirac cone structure at the K point within ±0.8 eV around the Fermi energy (E_F) (see Figure S2a). In a simple visual perspective of the density of states (DOS) (Figure S1c) as a function of the energy, one can notice that the two peaks closest to E_F correspond to the π and π^* flat band dispersion at the M point. The other intense peaks are not relevant to the present study since our work is restricted to the Dirac cone. Accordingly, the Monkhorst–Pack grid of 720 × 720 × 1 allows for estimating the limits of Dirac cone approximation and the average charge carrier velocity (v_F, group velocity) of graphene. The latter is the key input parameter of the semi-analytical modeling approach.

By using the high-resolution Monkhorst–Pack grid, the group velocity of graphene was found to be $v_F = 0.829 \times 10^6$ m s^{-1}, which is the average value obtained by linear fitting of the highest valence band ($v_{F-\pi} = 0.827 \times 10^6$ m s^{-1}) and the lowest conduction band ($v_{F-\pi^*} = 0.832 \times 10^6$ m s^{-1}) concerning the magnitude of the wave vector around the K point (Figure S2a). Figure S2b confirms the (quasi) linear variation of the group velocity of graphene for selected values of energy band dispersion. It should be noted that for large values of the wave vector, the band dispersions are anisotropic as a result of the flat band dispersion of the π and π^* bands (±2 eV).

Even though, if the value of the wave vector gets closer to the dispersion at the K point, the energy band dispersion of the π band (Figure S2c) and π^* band (Figure S2d) deviates slightly from a linear behavior of charge carrier velocity (dashed rectangular regions). This significant finding displays the limit of applicability of the present modeling approach to study, i.e., the semi-analytical model works in an energy range of ±0.2 eV or about 50 THz.

2.2. Semi-Analytical Model: Estimating the Electronic Properties

As proposed by Popov et al. [33], the linear band dispersion of charge carriers in graphene (near the Fermi energy and around the K point) is the foundation of the semi-analytical model. The group velocities of the π electrons and π^* holes can be obtained by linear fitting of the related band dispersions vs. the crystal momentum (discussed in Section 2.1). However, in the Dirac cone approximation, the electron band dispersion (E) is determined as [40]:

$$E = \pm v_F |\mathbf{p}| \tag{2}$$

where the ±sign represents the π^*—upper band and π—lower band, and $\mathbf{p}$ is the crystal momentum denoted as $\mathbf{p} = \hbar(\mathbf{k} - \mathbf{\Gamma K})$ ($\hbar$ is the reduced Planck constant).

In contrast, in GNRs, several (sub)bands (E_n) appear because the quasi-one-dimensional charge carrier confinement in a rectangular (thin) strip whose band structure is described by [41,42]:

$$E_n = \pm\frac{\Delta}{2}\sqrt{n^2 + \frac{2p_{\parallel}^2}{m^*\Delta}} \tag{3}$$

where Δ is the corresponding bandgap, n is the 1D band-index number ($n = 1, 2, 3, \ldots$) $p_{\parallel}$ is the parallel wave vector to the direction of the GNR, and m^* remains as the effective electron mass.

It is worth noting that Equation (3) illustrates, at the Γ point in GNRs, a parabolic band dispersion in the eV scale for narrow-width GNRs (e.g., $\leq$10 nm) (Figure S1e) whereas a linear and (quasi) gapless band dispersion is observed for sufficiently wide-width GNRs (e.g., $\geq$100 nm) (Figure S1f). As the ribbon width increases ($w \to \infty$) the structural conformation and band structure of graphene are recovered. Therefore, wide enough GNRs (Figure S1d) are expected to display graphene-like properties. It is important to note that, in wider GNRs, the effect of the basal plane is more significant than the effect of the edges, while in narrower GNRs, the edges (specifically, whether they are armchair or zigzag edges) are more prominent. In particular, Yang et al. [17] have experimentally shown that the transport properties of graphene are significantly impacted when the ribbon width is $\leq$57 nm. In the present study, we examine systems that are up to eight times wider than this threshold, ensuring the validity of the proposed model.

The bandgap of GNRs can be calculated by using the group velocity of graphene and the ribbon width (w) as the input. The equation for this calculation is given as:

$$\Delta = \frac{2\pi v_F \hbar}{w} \tag{4}$$

Otherwise, the effective electron mass can be determined by using the estimated bandgap and group velocity of graphene, according to the following equation:

$$m^* = \frac{\Delta}{2v_F^2} \tag{5}$$

2.3. Semi-Analytical Model: Describing the Plasmonic Properties

It has been observed experimentally that, in 2D GNR arrays, the plasmon wavelength follows the sample length, which is much larger than the distance between adjacent ribbons and the widths [18]. Interestingly, even though single GNRs can be asymmetric, 2D GNR arrays that are wide enough can be viewed as well-defined 2D planes where the group velocity of graphene can be used to study the plasmonic properties of these systems. With this in mind, the following approach can be used to obtain the plasmon frequency-momentum dispersion [33]:

$$\omega = \mathrm{Re}\left[\sqrt{\frac{2\pi e^2 N_{2D}}{\epsilon m^*} q\cos^2\theta - \frac{v^2}{4}} - i\frac{v}{2}\right] \tag{6}$$

- e is the well-known electron charge,
- N_{2D} is the 2D electron density,
- ϵ is the dielectric constant,
- q represents the reciprocal wave vector,
- θ represents the plasmon excitation angle,
- v is the electron relaxation rate.

Additionally, if 2D GNR arrays are unsupported (freestanding case) or supported (non-freestanding case), the plasmonic properties can be described by recalculating the effective electron mass (m^*) with the corresponding group velocity of graphene, as reported in [43].

Otherwise, the parameters of Equation (6), i.e., θ, v, and N_{2D}, can be easily manipulated by changing their values or adapting experimentally. From a theoretical notion or experimental side, N_{2D} can also be extracted from the concept of the Fermi level (E_F) [44], as follows:

$$E_F = \hbar v_F \sqrt{2\pi N_{2D}} \tag{7}$$

now, for an intrinsic semiconductor, E_F is found in the middle of the bandgap, i.e., the number of electrons and holes are the same, giving the following expression:

$$E_F = \frac{E_c + E_v}{2} + \frac{k_B T}{2} \ln\left(\frac{N_v}{N_c}\right) \tag{8}$$

where N_c is the effective DOS in the conduction band, N_v is the effective DOS in the valence band, E_c is the conduction band edge, E_v is the valence band edge, k_B is Boltzmann's constant, and T is the absolute temperature.

Hereafter, by the combination of Equations (7) and (8), Equation (6) can be expressed in terms of the needed Fermi level shift as well as the desired temperature. In the present case, we focus only on 2D electron density (N_{2D}) values that are experimentally feasible [33,45]. The equations described so far (Equations (2)–(8)) show a clear dependence on the group velocity of graphene, which cannot be chosen arbitrarily (e.g., $v_F \sim 10^6$ m s^{-1}), but must be accurately determined (e.g., by DFT computations as in this work) or measured experimentally.

It is necessary to emphasize that for large values of damping, the radicand in Equation (6) becomes negative, i.e., surface plasmons are absent or shifted to large momenta. Hence, this effect can be understood in terms of the complex (macroscopic) dielectric function [46] denoted as:

$$\epsilon_M(\mathbf{q}, \omega) = \frac{1}{\epsilon^{-1}(\mathbf{q}, \omega)} \tag{9}$$

Equation (9) states that the existence of a plasmon is determined by zeros in the real part of the frequency range where the imaginary part is small or zero. If this condition is not met, only single-particle excitations can be found instead of collective excitations.

2.4. Semi-Analytical Model: Showing the Plasmon Spectrum

Lastly, the plasmon spectrum (the maximum of the plasmon peak) for selected q values can be attained by means of the conventional approach of the spectral line profile using a Lorentzian function, as follows [47]:

$$L = \frac{1}{1 + x^2} \tag{10}$$

where, L is the standard Lorentzian function fixed to a maximum value of 1, and x is an auxiliary (dimensionless) variable denoted as:

$$x = \frac{2(\omega - \omega_0)}{W} \tag{11}$$

here, ω_0 represents the transition frequency of the maximum (THz), ω is the frequency (energy) range of interest, and W is the full width at half maximum (FWHM). The W parameter was set to 0.5 for all spectra unless otherwise stated.

At that point, the plasmon spectrum can be plotted by rearranging Equations (10) and (11) as:

$$L = \frac{1}{1 + \frac{4(\omega - \omega_0)^2}{W^2}} \tag{12}$$

Four important facts should be highlighted:

(i) Lorentzian line function describes the form of a spectroscopical feature corresponding to a frequency (energy) change in ions, molecules, atoms, or now 2D GNR arrays [48].

(ii) The units of ω, ω_0, and W are typically wavenumbers (for absorption spectra) or frequency (for lifetime spectrum of collective excitations) [49]. In the present study, the last entry is used, that is, in terms of frequency, in what we call the plasmon spectrum or maximum of the plasmon peak.

(iii) The maximum of the plasmon peak could be delayed due to core-electron excitations which are not taken into account in the Lorentzian function [50].

(iv) The plasmon structure is not expected to be a straightforward Lorentzian peak, but the approach used is useful in clearly showing the control of surface plasmons in 2D GNR arrays.

3. Results and Discussions

3.1. Electronic Properties of Freestanding GNRs

Based on Equation (4) and $v_F = 0.829 \times 10^6$ m s^{-1}, Figure 1 shows the bandgap (Δ) as a function of ribbon width ranging from 0.4 to 2.7 nm (ultra-narrow GNRs, Figure 1a) and from 15 to 90 nm (wide GNRs, Figure 1b). The prediction given from 1 to 6 eV by the semi-analytical model (red curve) is in agreement with the predictions of the sophisticated GW approximation for different types of GNRs (3p+1, 3p, and 3p+2) [51]. It is important to keep in mind that the edges (armchair or zigzag edges) of the nanoribbon play a crucial role in determining the bandgap for ultra-narrow GNRs. These edge effects cannot be taken into account explicitly by the simplified model of Equation (4), but the semi-analytical model provides a reasonable prediction for nanoribbons greater than 1 nm, particularly for 3p+1 and 3p families. As mentioned above, edge (atomistic) effects start to be negligible as the ribbon width increases (>60 nm [17]), going from an ultra-confinement on very narrow ribbons to a quasi-confinement on sufficiently wide ribbons. This statement is further confirmed since the bandgap changes dramatically from the eV scale down to a few meV, as discussed below.

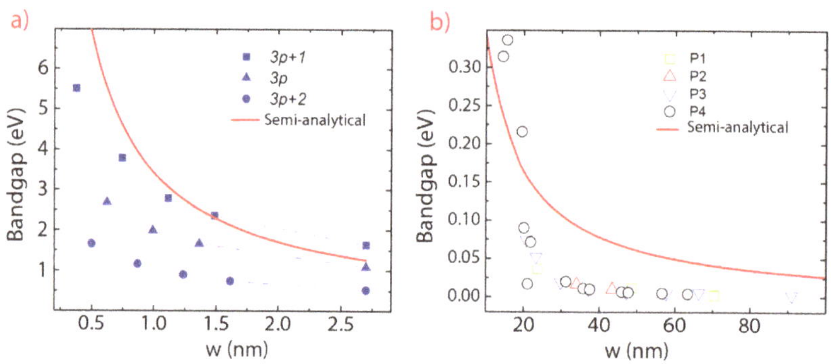

Figure 1. Bandgap vs. the ribbon width. The semi-analytical approach is compared with the results obtained by (**a**) the GW approximation for diverse types of GNRs and (**b**) experimental results (P1–P4 denotes different datasets).

Figure 1b compares the prediction of the semi-analytical model (red curve) with the experimental bandgap values (<300 meV) of different GNRs (P1–P4 refers to four different datasets [42]). The predicted bandgaps are consistent across GNRs with widths down to 20 nm. From 20 to 40 nm, the smooth prediction of the semi-analytical model cannot capture the steep dropoff of about 60 meV compared to experiments. Interestingly, the dropoff is reduced to ~25 meV from 40 to 70 nm and as low as 10 meV at 90 nm. While the semi-analytical model may not perfectly match experimental results, it is important to stress that it is not practical to use ab initio methods for electronic properties of wide nanoribbons, as their widths are in the range of tens of nanometers. Additionally, the discrepancy between the model and experiments is expected to be minimal (or non-existent) for experimental nanoribbons (≥155 nm), making the present modeling approach suitable for the GNR systems reported in Ref. [18].

The bandgap of experimental GNRs with widths of 155, 270, 380, and 480 nm are shown in Figure 2 and Table 1. The bandgaps are found to be very small, in the range of a few meV, and the effective electron masses are in agreement with previous studies (Table 1) [42,51]. The semi-analytical model predicts a decrease in bandgap as the ribbon width increases, which is consistent with the experimental results and earlier theoretical reports.

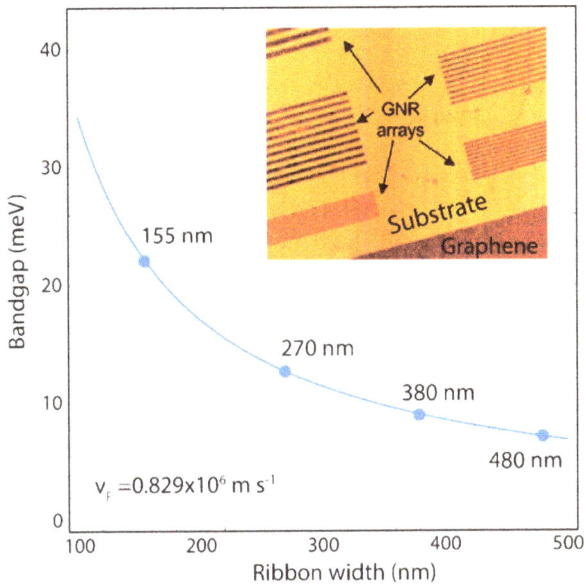

Figure 2. Bandgap vs. the ribbon width of experimental GNRs. The blue line represents a prediction for bandgaps in GNRs with widths ranging from 100 to 500 nm. Figure inset shows 2D periodic arrays of experimental GNRs (adapted with permission of Ref. [18]).

Table 1. Bandgap and electron effective mass of experimental GNRs estimated by the semi-analytical model. The free-electron mass is symbolized as m_0.

Ribbon Width (nm)	Bandgap (meV)	Effective Mass (m^*)$\times m_0$
155	22.12	2.83×10^{-3}
270	12.70	1.63×10^{-3}
380	9.02	1.16×10^{-3}
480	7.14	0.91×10^{-3}

Figure S3 shows the band structure and DOS of experimental GNRs. The band structure is obtained by applying Equation (3) with the bandgap and effective mass values shown in Table 1. Nine sub-bands ($n = 9$) below/above the Fermi level are introduced, i.e., nine valence and nine conduction sub-bands. It is worth mentioning that the band structure for each graphene nanoribbon is constructed using the same effective mass for all bands. However, when examining plasmonic properties, only the first conduction band and last valence band ($n = 1$) are considered.

The band structure and DOS of the experimental GNRs differ from those of graphene, regardless of the ribbon width. Specifically, a bandgap opening in a few meV is observed at the Γ point, with a parabolic-like band dispersion on a small energy scale and strong peaks in the DOS near the Fermi level. In contrast, graphene has a linear band dispersion at the

K point and no DOS at the Fermi level (Figure S1b,c). In GNRs, this bandgap opening is a direct result of the quasi-one-dimensional confinement of charge carriers, which makes all GNRs semiconducting materials. As stated, the systems under study have very small bandgaps (Table 1), suggesting that they could have plasmonic responses in the terahertz range, which indeed is the range of many photonic and plasmonic applications [52].

Otherwise, the current semiconductor technology requires bandgaps around 1 eV (such as silicon ~1.1 eV), in which case, narrow-wide GNRs are the best choice. As an example, using Equation (4), $v_F = 0.829 \times 10^6$ m s^{-1} (freestanding case), and $w = 2.7$ nm (experimentally observed width in GNRs on Ge(001) [53]), the corresponding bandgap is 1.27 eV, slightly above the experimental value (~0.9 eV) detected by scanning tunneling spectroscopy (STS) measurements [53].

It is well known that the group velocity of graphene is reduced up to 40% by interacting with germanium or other substrates (non-freestanding case) [15,43]. Keeping this in mind, Figure 3 displays the band structure of experimentally realized GNR on Ge(001) (2.7 nm wide) for different percentage decreases in the group velocity of graphene estimated by LDA-DFT ($v_F = 0.829 \times 10^6$ m s^{-1}) (see Table 2). Surprisingly, a perfect agreement between the theoretical and experiment bandgap can be observed at a 30% decrease in the group velocity, i.e., $\Delta_{30\%} = 0.88$ eV.

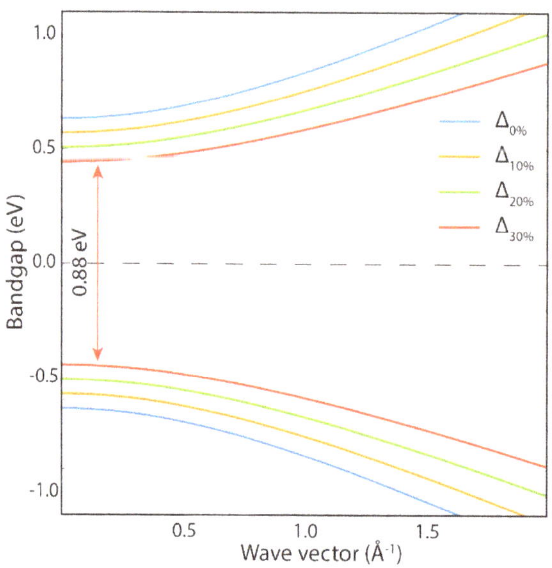

Figure 3. Band structure of experimental realized GNR on Ge(001), considering different percentage decreases (0%, 10%, 20%, 30%) in the group velocity of graphene. The parameters were fixed as: $w = 2.7$ nm and $n = 1$.

Table 2. Bandgap of experimentally realized GNR of 2.7 nm width as a function of the percentage decrease in the group velocity.

v_F	v_F	**Bandgap (meV)**
0	0.829	1.27
10	0.746	1.14
20	0.663	1.02
30	0.580	0.89
40	0.497	0.76

Furthermore, Ruffieux et al. [54] have reported the fabrication of armchair graphene nanoribbons about 1.5 nm wide on Au(111) with a bandgap of 2.3 eV, an effective electron mass of 0.21 m_0, and group velocity of 0.82×10^6 m s^{-1}. These results were obtained by angle-resolved photoelectron spectroscopy and scanning tunneling spectroscopy. Interestingly, using the experimental bandgap and ribbon width, the semi-analytical model estimates values practically equal to those observed experimentally ($v_F = 0.809 \times 10^6$ m s^{-1} and $m^* = 0.27 m_0$).

The agreement between the experiments and predictions, despite not explicitly taking into account edge effects or other factors (such as passivation, structural disorder, or charge at the edges), confirms that the semi-analytic model is suitable for describing the electronic properties of GNRs.

3.2. Electronic Properties of Non-Freestanding GNRs

As demonstrated, Equation (4) can be used to determine the bandgap of wide nanoribbons on various substrates by changing the substrate and using the corresponding group velocity of graphene, as discussed in the previous section. Indeed, Hwang C. et al., [43] measured the group velocity of graphene on quartz (G/Quartz), hexagonal boron nitride (G/BN), and silicon carbide (G/SiC). Another piece of evidence came from cyclotron mass in graphene device (G/SG) [55], massless relativistic particles (Popov) [33], and LDA-DFT limit (DFT). In this context, we also suggest the growth of experimental (wide) GNRs on this type of substrates, whose predicted bandgap values are shown in Figure 4. Regardless of the substrate, the bandgap values are less than 100 meV and these values decrease as the ribbon gets wider.

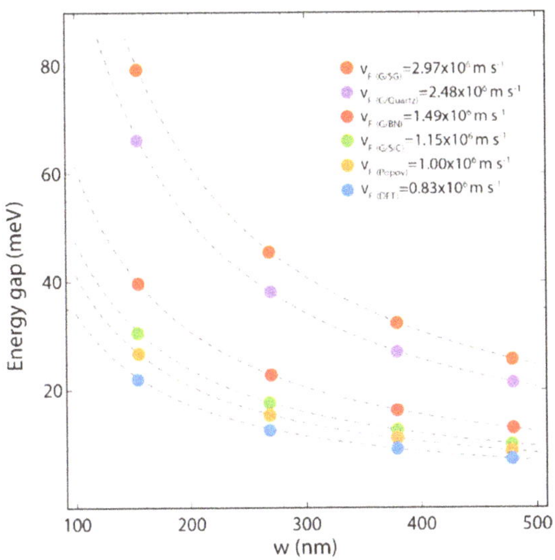

Figure 4. Bandgap variation as a function of the ribbon width. Experimental GNRs w = 155, 270, 380, 480 nm) are considered taking into account the group velocity of graphene on various substrates. Dashed curves indicate the bandgap for GNRs from 100 to 500 nm wide.

Interestingly, for high values of v_F, a large bandgap opening is observed, particularly, in GNRs of 155 nm wide. Such bandgap opening corresponds to a 71% increase in quartz, 50% in boron nitride, and 33% in silicon carbide, from the related values predicted by using LDA-DFT computations. Even an increase of 20% is observed compared to the massless relativistic approach. These results constitute the first observation of the dependence on the charge carrier velocity at a fixed ribbon width. This dependence allows achieving, by a smart choice of supporting substrate, the desired bandgap for a particular application in graphene-based electronic devices.

3.3. Comparison of Semi-Analytical Model with Previous Reports

Before delving into the specific plasmonic properties of the nanoribbons (freestanding and non-freestanding), we first demonstrate the effectiveness of the proposed model by comparing it to previous studies. Unfortunately, there is a lack of experimental evidence for plasmonic properties in the THz frequency range considered here (<50 THz), the 2D GNR arrays under study, and even more importantly in the reciprocal space. To overcome this limitation, we compare the semi-analytical model with more sophisticated modeling approaches, such as TDDFT+RPA (Figure S4) [19] and finite-difference time-domain (FDTD) (Figure 5b) [56], which are commonly used and have demonstrated outstanding agreement between theory and experiment. In addition, the semi-analytical model is tested with results reported for graphene microribbons (Figure 5a) [57].

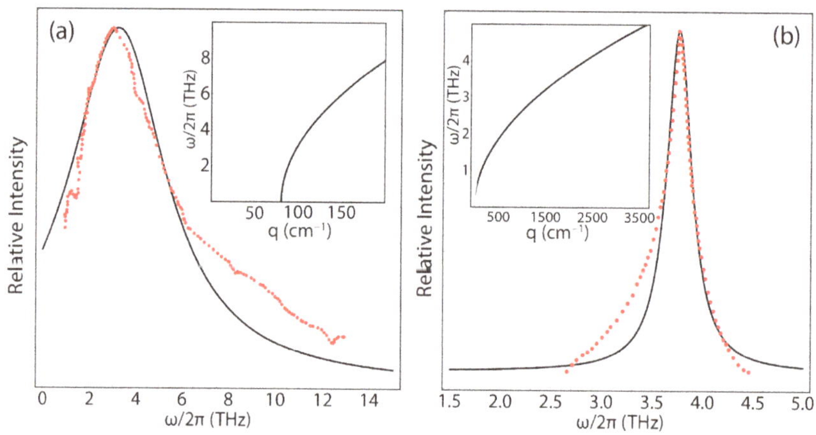

Figure 5. (**a**) Maximum of the plasmon peak ($\leq$15 THz) at momentum q = 100.6 cm^{-1}. The semi-analytical model (black curve) is compared with the data of Ref. [57] (red points) for graphene microribbon arrays of 4 μm wide. Figure inset shows the entire plasmon frequency dispersion up to q = 200 cm^{-1}. (**b**) Maximum of the plasmon peak ($\leq$5 THz) at momentum q = 2050 cm^{-1}. The semi-analytical model (black curve) is compared with the predictions of Ref. [56] (red points) for GNR arrays of 10 nm wide. Figure inset shows the entire plasmon frequency dispersion up to q = 3500 cm^{-1}.

Interestingly enough, Ju et al. [57] have explored the plasmon excitations in engineered graphene microribbon arrays deposited on Si/SiO_2, demonstrating the tunability of plasmon resonances over a broad THz frequency range by changing the ribbon width and electrostatic doping. For example, for one of the scrutinized systems with the following parameters reported: $w = 4$ μm, $q = 100.06$ cm^{-1}, $m^* \approx 1.3 \times 10^{-4} m_0$, a doping concentration of $N_{2D} = 1.5 \times 10^{13}$cm^{-2}, and electron mobility of 1000 cm^{-2}/V s ($\nu = 8.71 \times 10^{13}$ s^{-1}); the maximum of the plasmon peak is found at 3 THz. The semi-analytical model has accurately reproduced this result with a minor deviation of only 0.20 THz within the frequency range of 1 to 6 THz (Figure 5a, the spectra were normalized by the respective peak values for convenience of comparison). Moreover, the starting plasmon frequency dispersion is shifted to larger values of momenta ($q > 80$ cm^{-1}) (see Figure 5a inset).

On the other hand, in Ref. [56], the authors propose the fabrication of 2D arrays of GNRs of 10 nm wide on Si substrates for THz polarizer surface plasmons by using viable conditions such as: $m^* = 0.28 m_0$, $N_{2D} = 4.33 \times 10^{13}$ cm^{-2}, and assuming the maximum electron mobility ($\nu = 0$). The plasmon response predicted by FDTD was found in the frequency range from 1.81 to 3.81 THz with the maximum of the plasmon peak at 3.75 THz. By setting these parameters in the semi-analytical model, the plasmon spectrum predicted by the semi-analytical model is found in excellent agreement at $q = 2050$ cm^{-2} (Figure 5b, the spectra were normalized using the corresponding maximum values to facilitate comparison). Forbidden plasmon regions are absent in the plasmon frequency dispersion (Figure 5b inset).

From the ab initio side, in Ref. [19], the authors reported that freestanding and narrow armchair GNRs arranged as 2D periodic arrays have an effective electron mass of $0.36m_0$. By directly establishing this value in Equation (6), which in fact is independent of the group velocity of graphene at that point, a perfect agreement between both approaches can be observed in the plasmon frequency-momentum dispersion for a change in the Fermi level (doping) of up to 0.3 eV ($N_{2D} = 3.67 \times 10^{12}$ cm^{-2}, orange curve). For higher values of doping (0.4 eV, purple line), a discrepancy is observed. All these results allow us to demonstrate the reliability of the semi-analytical model and can be used to describe the plasmonic properties of 2D GNR arrays of interest.

A final remark, while the semi-analytical model accurately predicts the experimental results, it is important to note two important facts: (i) in practice, the non-freestanding GNRs will be supported by two anchors and may vibrate mechanically, altering their phonon structure, and (ii) nearby materials could affect the electric or magnetic fields generated. Hence, the simplified model does not take into account these factors, so caution should be taken during experiments to avoid inconsistencies. This can be achieved through proper gating voltage selection or adjusting the electric field applied to ω, which is given as $E(\omega) = E_0\exp(-i\omega t)$.

3.4. Plasmonic Properties of Non-Freestanding GNRs

Now, Figure 6 shows the plasmon frequency-momentum dispersions of 2D GNR arrays from 155 to 480 nm and subject to different group velocities. The parameter of Equation (6) ($N_{2D} = 1.0 \times 10^{12}$ cm^{-2}, $\nu = 1.0 \times 10^{13}$ s^{-1}, and $\theta = 0$) have been fixed based on experimentally feasible values [33,45]. As observed, the plasmon trends follow a $\sqrt{q}$—like dispersion from 0 to 10,000 cm^{-1}. This plasmon dispersion is characteristic of 2D materials [55]. More importantly, by increasing the value of v_F, the plasmon frequency increases.

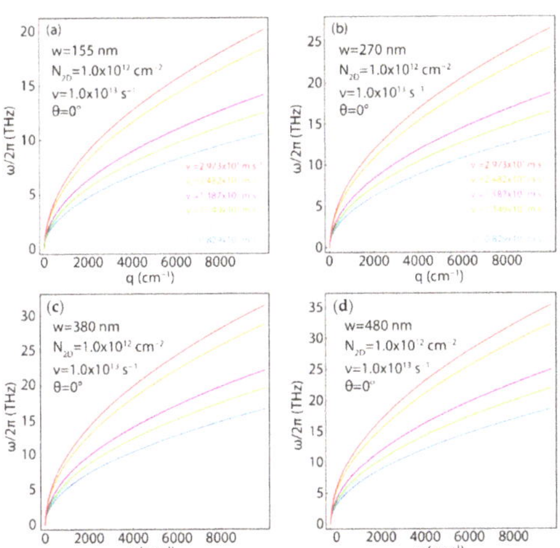

Figure 6. Frequency–momentum dispersion as a function of the plasmon momentum, considering the group velocity of graphene on various substrates for 2D GNR arrays of (**a**) $w = 155$ nm, (**b**) $w = 270$ nm, (**c**) $w = 380$ nm, (**d**) $w = 480$ nm.

For an illustrative example, at the end of the sampled momentum ($q = 10,000$ cm^{-1}), an increase of ~9 THz is observed for 2D GNR arrays of 155 nm wide (Figure 6a), considering the group velocity of graphene obtained by LDA-DFT computations ($v_F = 0.829 \times 10^6$ m s^{-1}, blue curve) and that obtained from graphene on quartz ($v_F = 2.482 \times 10^6$ m s^{-1}, orange curve). Similarly, an increase of almost 13 THz is detected for 2D GNR arrays of 270 nm wide (Figure 6b), 16 THz for 2D GNR arrays of 380 nm wide (Figure 6c), and 20 THz for 2D GNR arrays of 480 nm wide (Figure 7d).

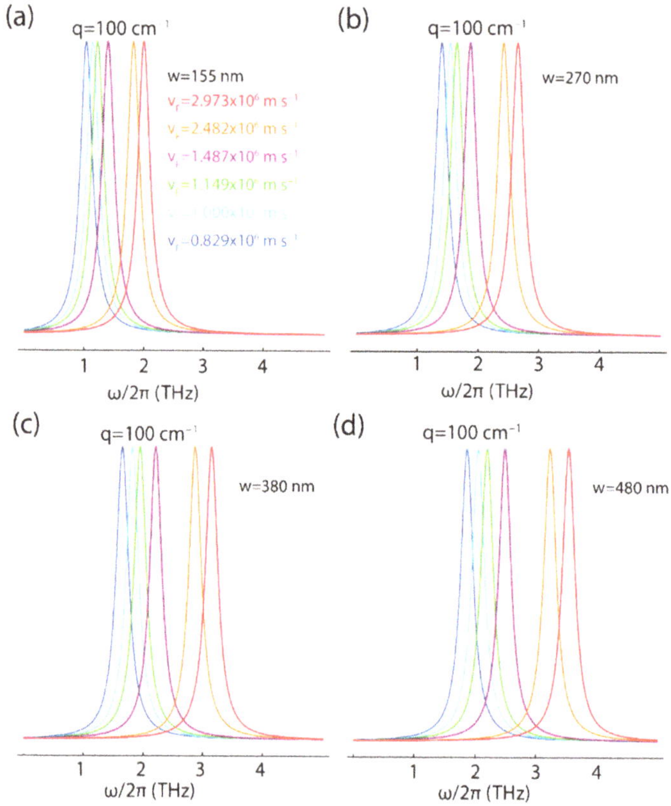

Figure 7. Maximum of the plasmon peak (≤ 5 THz) at plasmon momentum $q = 100$ cm^{-1}, considering the group velocity of graphene onto various substrates for 2D GNR arrays of (**a**) $w = 155$ nm, (**b**) $w = 270$ nm, (**c**) $w = 380$ nm, (**d**) $w = 480$ nm.

By implementing Equation (12) at $q = 100$ cm^{-1}, Figure 7 demonstrates the controllability and tunability of the plasmon response of 2D GNR arrays by varying the group velocity of graphene as well as the ribbon width: $w = 155$ nm (Figure 7a), $w = 270$ nm (Figure 7b), $w = 380$ nm (Figure 7c), and $w = 480$ nm (Figure 7d). Interestingly, it can be seen that keeping momentum fixed (i.e., $q = 100$ cm^{-1}) while varying the ribbon width and the charge carrier velocity, the maximum of the plasmon peak is found at higher frequencies, suggesting that the effect of group velocity is more critical in wider nanoribbons. To clarify this fact, for $w = 155$ nm and freestanding cases (blue curves), the maximum of the peak is almost at 0.9 THz (Figure 7a) while for $w = 480$ nm, the maximum of the peak is observed at almost 1.4 THz (Figure 7d). On the contrary, for graphene on quartz (orange curves), the maximum of the plasmon peak is observed at almost 2 THz (Figure 7a) for $w = 155$ nm, and at almost 3.5 THz for $w = 480$ nm (Figure 7d). From these results, it is noticed a percentage variation of up to 35% and 43% in the position of the maximum of the plasmon peak, in each case respectively.

3.5. Plasmonic Properties of Freestanding GNRs: The Effect of Ribbon Width (w)

The results discussed in the previous sections allow us to understand the versatility of 2D periodic arrangements of GNRs (i.e., 2D GNR arrays) by changing the supporting substrate and ribbon width, which can be exploited in the design of novel nanophotonic devices.

In what follows in the article, we focus on the freestanding cases and parameters such as 2D electron concentration (doping) (N_{2D}). While the parameters of Equation (6) can be varied trivially, the goal is to be able to use this simple model with parameters that can be fixed and measured in experiments. In this context:

- The manipulation of the plasmon excitation angle (θ) in 2D GNR arrays is imperative to obtain the surface plasmon resonance and the related sensitivity. However, Equation (6) must be used with caution since, at $\theta = 90$, the radicand becomes zero. More importantly, the properties and features of oblique plasmons can be scrutinized.
- In conductivity-related phenomena, charge carrier mobility plays an important role in the electronic and plasmonics properties of any system. Thus, high-charge carrier mobility is commonly observed in defect-free graphene samples, whereas charge carrier mobility is significantly reduced in graphene samples with high concentrations of defects. From the experimental part, this effect depends on the synthesis process. From the theoretical part, in Equation (6), this fact is introduced by the electron relaxation rate (ν), taking into account the following statement: a high value of ν is connected with a low charge carrier mobility, and vice-versa.
- 2D electron concentration (N_{2D}) (or 2D charge carrier density) can be controlled by injecting or ejecting electrons, i.e., by either doping the GNR system or by gating voltages. The latter technique is of particular interest because it can be varied in a reasonable range up to $N_{2D} = 5.0 \times 10^{12}$ cm^{-2} [42].

Before discussing the effect of the 2D electron concentration (Section 3.6), we probe a little into the effect of ribbon width in freestanding cases. Figure S5a displays the plasmon frequency–momentum dispersion of 2D GNR arrays for $w = 155$ nm (blue curve), $w = 270$ nm (cyan curve), $w = 380$ nm (green curve), and $w = 480$ nm (red curve). Again, the important fact is the increase in the plasmon frequency with increasing ribbon width. To further evidence the tunability of plasmon response, Figure S5b shows the maximum of the plasmon peak at $q = 1000$ cm^{-1}. Specifically, the peak position is found at 3.35 THz for $w = 155$ nm, at 4.34 THz for $w = 270$ nm, at 5.26 THz for $w = 380$ nm, and at 5.91 THz for $w = 480$ nm.

Figure S5c shows the percentage variation in plasmon frequency for three different testing momenta ($q = 100, 1000, 10,000$ cm^{-1}). Although it is observed that increasing the ribbon width also increases the plasmon frequency, it can be noted that this effect is independent of momentum, resulting in a constant trend in plasmon frequency rate in all cases, that is, ~25% from 155 to 270 nm, ~16% from 270 to 380 nm, and ~11% from 380 to 480 nm. A close view of Figure S5c shows a slight variation at 100 (black marker), 1000 (red marker), and 10,000 (blue marker) cm^{-1}, because the width increment from 155 to 270 nm results in a percentage variation of 25.36%, 24.38%, and 26.16%, respectively. Additionally, it can be noticed that the plasmon frequency rate can be described with a decreasing exponential function, which suggests that as the ribbon width increases, the effect on the plasmon frequency dispersion is gradually minor. The decreasing behavior of the plasmon frequency rate can be corroborated by inspecting more values of q.

3.6. Plasmonic Properties of Freestanding GNRs: The Effect of 2D Charge Density (N_{2D})

The parameters in Equation (6) can be easily adjusted by any researcher, but the goal is to use the model with experimental results or those obtained from more advanced predictions (such as TDDFT + RPA [19] and FDTD [56]). This study aims to provide a practical guide for describing the sensitivity and detection of surface plasmons in real-world situations, at least in rough approximation.

Based on the measurements of Ref. [45], three different 2D charge carrier concentrations are used ($N_{2D} = 1.0 \times 10^{12}$ cm^{-2}, $N_{2D} = 2.0 \times 10^{12}$ cm^{-2}, and $N_{2D} = 4.0 \times 10^{12}$ cm^{-2}) (Figure 8a–d). Regardless of the ribbon width, increasing the value of N_{2D} also significantly increases the plasmon frequency. This plasmon frequency increases by up to almost 40 THz (Figure 8d) which is much greater than that observed if only the ribbon width is increased (Figure S5a), i.e., approximately twice in magnitude. The entire frequency-momentum dispersion is below 21 THz for $w = 155$ nm (Figure 8a), 27 THz for $w = 270$ nm (Figure 8b), 33 THz for $w = 380$ nm (Figure 8c), and 40 THz for $w = 480$ nm (Figure 8d). Note that the largest increase in plasmon frequency is achieved at $N_{2D} = 4.0 \times 10^{12}$ cm^{-2} (red curves). It is also detected that the plasmon dispersion starts at zero frequency and zero momentum, that is, forbidden plasmon regions are unexpected to occur. If the plasmon response needs to be shifted to large values of q, high values of plasmon excitation angle or electron relaxation rate can be introduced (separately or combined) in Equation (6), for instance, at $\theta = 80$ and $\nu = 4.0 \times 10^{13}$ s^{-1}.

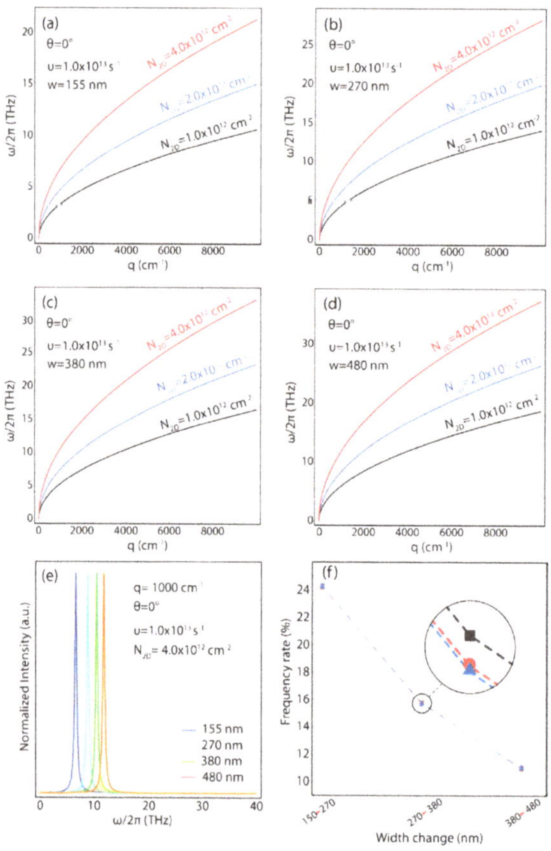

Figure 8. Frequency–momentum dispersion as a function of the plasmon momentum using the charge carrier velocity of freestanding graphene ($v_F = 0.829 \times 10^6$ m s^{-1}) for the experimental 2D GNR arrays of (**a**) $w = 155$ nm, (**b**) $w = 270$ nm, (**c**) $w = 380$ nm, (**d**) $w = 480$ nm. (**e**) Maximum of the plasmon peak ($\leq$40 THz) at plasmon momentum $q = 1000$ cm^{-1}, considering different ribbon widths. (**f**) Percentage variation in plasmon frequency rate as a function of the ribbon width increment, for three selected plasmon momenta ($q = 1000$, 5000, 10,000 cm^{-1}).

Figure 8e clearly shows the controllability and tunability of the maximum of the plasmon peak for a fixed momentum ($q = 1000$ cm^{-1}) and at the highest value of 2D charge carrier concentration ($N_{2D} = 4.0 \times 10^{12}$ cm^{-2}). The maximum is found at 6.72 THz for $w = 155$ nm (blue curve), at 8.87 THz for $w = 270$ nm (cyan curve), at 10.53 THz for $w = 380$ nm (green curve), and at 11.83 THz for $w = 480$ nm (orange curve). In general, the plasmon frequency end is detected at 21.26 THz ($w = 155$ nm), 28.06 THz ($w = 270$ nm), 33.29 THz ($w = 380$ nm), and 37.42 THz ($w = 480$ nm). These results fall below the limit of the semi-analytical model used here and are anticipated to match future experimental findings.

Figure 8f shows that the plasmon frequency rate is not affected by the variation of the ribbon width for testing momenta ($q = 1000, 5000, 10,000$ cm^{-1}). As result, one can note a constant trend in the plasmon frequency rate in all cases, that is, ~24% from 155 to 270 nm, ~16% from 270 to 380 nm, and ~11% from 380 to 480 nm. Additionally, it has been confirmed that the plasmon frequency rate can be represented by a decreasing exponential function, indicating that as the ribbon width increases, its impact on the plasmon frequency dispersion becomes progressively smaller.

4. Chlorpyrifos-Methyl Sensing

We now explore the potential application of 2D GNR arrays in molecular sensing (Figures 9 and 10 and Table 3). Chlorpyrifos-methyl is an insecticide that is widely used to control insect pests on various crops, but it is highly toxic to organisms and humans. When exposed to it, negative impacts on the nervous system can occur, including headaches, dizziness, and nausea. Research has connected chlorpyrifos-methyl to developmental problems in children, such as decreased IQ and attention deficit disorders. Given these concerns, numerous nations have taken action to restrict or prohibit the utilization of chlorpyrifos-methyl [58,59]. Therefore, effective detection and removal methods are crucial. From a chemical perspective, chlorpyrifos-methyl has a benzene ring with perpendicular π electrons that are likely to interact directly with the perpendicular π electrons of the GNRs via $\pi - \pi$ interactions. Additionally, this molecule has a resonance plasmon peak at 0.95 THz [60]. These types of molecules are often detected in water [59], but water has a strong THz absorption at 1 THz [61]. In this context, materials with resonance peaks within the range of water frequency absorption significantly enhance the interaction between incident THz waves and target molecules, making them useful for sensing proteins, nucleic acids, and cells.

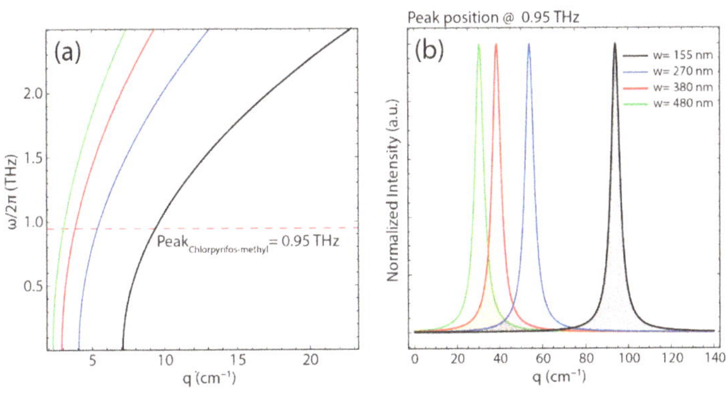

Figure 9. (**a**) Frequency–momentum dispersion as a function of the plasmon momentum using the charge carrier velocity of freestanding graphene ($v_F = 0.829 \times 10^6$ m s^{-1}) for the experimental 2D GNR arrays ($w = 155, 270, 380, 480$ nm). (**b**) Maximum of the plasmon peak at 0.95 THz as a function of the plasmon momentum for the same systems.

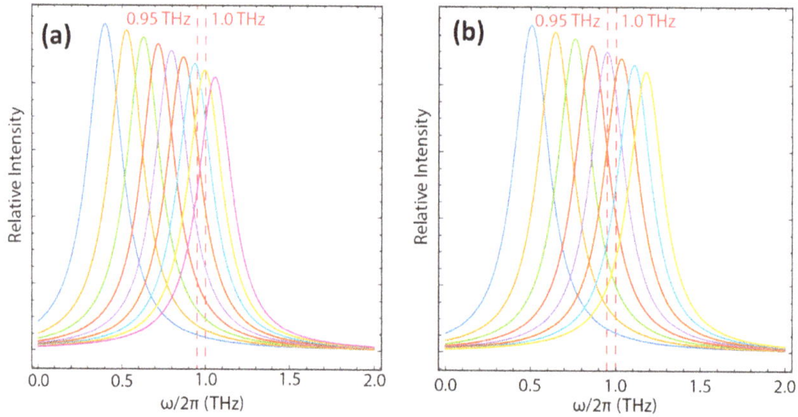

Figure 10. Dispersion of the maximum of the plasmon peak (<2 THz) with FWHM = 5 for (**a**) $w = 155$ nm by considering different momenta from 75 (blue curve) to 100 cm^{-1} (purple curve), and (**b**) $w = 480$ nm by considering different momenta from 25 (blue curve) to 35 cm^{-1} (orange curve).

Table 3. Position of the maximum of the plasmon peak at 0.95 THz for the experimental 2D GNR arrays analyzed.

Ribbon Width (nm)	Peak Position (cm^{-1})
155	94.2
270	54.1
380	38.5
480	30.5

Figure 9a shows that 2D GNR arrays can effectively serve as a sensing platform for detecting chlorpyrifos-methyl molecules by using a specific setting, for example: $N_{2D} = 2.5 \times 10^{12}$ cm^{-2}, $\theta = 0$, and $\nu = 2.24$ s^{-1}. All the analyzed cases ($w = 155$ nm, black curve; $w = 270$ nm, blue curve; $w = 380$ nm, red curve; $w = 480$ nm, green curve) have the necessary THz response. It is worth noting that high values of the electron relaxation rate (ν) and charge carrier concentration (N_{2D}) are used, which implies two important facts:

(i) The use of $\nu = 2.24$ s^{-1} is justified by the possibility of using GNRs with a high concentration of defects, such as oxidized graphene ribbons [62].
(ii) It is assumed that the interaction between the 2D GNR array and the chlorpyrifos-methyl molecule through $\pi - \pi$ interactions will cause a transfer of charge (doping) from the molecule, which validate the use of $N_{2D} = 2.5 \times 10^{12}$ cm^{-2}.

Additionally, in Figure 9a, small forbidden plasmon regions are detected at $q < 80$ cm^{-1} for $w = 155$ nm and at $q < 40$ cm^{-1} for $w = 270$ nm. In the remaining two cases ($w = 380$ nm and $w = 480$ nm), the forbidden regions appear at $q < 20$ cm^{-1}. Thus, as the ribbon width increases, the momentum range for which the plasmon does not exist shrinks, indicating that for $w \to \infty$, the plasmon dispersion resembles the results of 2D graphene (i.e., THz $= 0$ and $q = 0$). Additionally, the plasmon frequency dispersion increases as the ribbon width increases, reaching nearly 6 THz at $q = 300$ cm^{-1} and $w = 480$ nm (green curve, Figure S6). Importantly, Figure 9b shows that while all 2D GNR arrays have plasmon responses at 0.95 THz, these resonance modes shift to lower values of q as the ribbon width increases. The same shift has been observed in experimental graphene microribbons ranging from 1 to 4 μm [57]. For example, in Table 3, for $w = 155$ nm, the peak position at 0.95 THz is at $q \approx 94$ cm^{-1} (black curve) whereas for $w = 480$ nm the peak position is at $q \approx 30$ cm^{-1} (green curve). Using 2D GNR arrays with a larger range of ribbon widths, THz resonance from 0.1–1 THz can be straightforwardly engineered.

Subsequently, Figure 10 depicts the presence of plasmon resonances in 2D GNR arrays with widths of 155 nm (Figure 10a) and 480 nm (Figure 10b), respectively. These resonance modes are found in the same frequency of water (1 THz) at $q \approx 99$ cm^{-1} for $w = 155$ nm (orange curve) and $q \approx 30$ cm^{-1} for $w = 480$ nm (brown curve). In the context of plasmonic–organic molecule interactions, it is worth mentioning that the interaction between the plasmonic wave and organic molecules can cause changes in the signal that is generated. In particular, this interaction can cause changes in the resonance frequency of the plasmon modes detected in the analyzed 2D GNR arrays, leading to modifications in the optical signal observed here (for water or chlorpyrifos-methyl). The precise nature of these signal modifications depends on the specific details of the interaction between the plasmonic wave and the organic molecule, further including the size and shape of the molecule, its electrical properties, and its position relative to the incident plasmon wave. Understanding these interactions is an important area of plasmonics research and has immediate applications in fields such as biosensing, photonics, and optoelectronics as well as the design of novel graphene-based biosensors.

5. Conclusions

In summary, we used a semi-analytical approach, which takes into account the Dirac cone approximation and the group velocity of graphene ($v_F = 0.829 \times 10^6$ m s^{-1}) to examine the electronic and plasmonic features of GNRs with widths that match those of actual samples (155, 270, 380, and 480 nm). Both freestanding and non-freestanding models were considered. The band structure and bandgap were evaluated in the range of a few meV to eV. Plasmon resonances were evaluated in the frequency range below 50 THz.

From the viewpoint of electronic properties, this modeling approach reasonably reproduces the bandgaps of experimental samples (>20 nm wide) and values predicted by sophisticated GW approximation (with widths from 1.5 to 2.7 nm). Moreover, it is possible to match the experimental bandgap (~0.9 eV) of GNRs (10 nm wide) on Ge(001) at a 30% decrease in the group velocity of graphene estimated by LDA-DFT computations as well as the effective electron mass ($m^* = 0.27m_0$) of GNRs (~1 nm wide) on Au(111). With this evidence, this work demonstrated that the bandgaps of the systems under examination ($\geq$155 nm) range from 7 to 22 meV. These bandgaps values can be increased by increasing the group velocity of graphene when it is supported on various substrates such as quartz, and hexagonal boron nitride, among others.

Considering the plasmonics side, this work is restricted to the analysis of the effect of changing the group velocity of graphene, ribbon width, and charge carrier concentration (doping). Other parameters such as plasmon excitation angle, electron mobility, temperature, and Fermi level shift can be scrutinized by simple modification of these values in Equation (6). Specifically, the semi-analytical model reproduced the experiment of graphene microribbons of 4 μm wide organized as 2D periodic arrays, finding the maximum of the plasmon peak at 3.20 THz for $q = 100.06$ cm^{-1} as well as the predictions of GNRs of 10 nm wide on Si with the maximum of the plasmon peak at 3.75 THz for $q = 2050$ cm^{-1}. For the systems under study, the plasmon frequency dispersion rises as the group velocity of graphene increases. The plasmon frequency further increases by increasing the ribbon width or the charge carrier concentration up to about 50 THz for $q = 10,000$ cm^{-1}. The decay of the plasmon frequency rate suggested that as the ribbon width grows, the effect on the plasmon frequency dispersion became gradually minor.

The potential application of the systems considered was discussed in terms of molecular sensing of chlorpyrifos-methyl, which is a highly toxic pesticide for organisms and humans. By setting the following parameters: $N_{2D} = 2.5 \times 10^{12}$ cm^{-2}, $\theta = 0$, and $\nu = 2.24$ s^{-1}, all the analyzed cases displayed plasmons resonances at the same frequency of chlorpyrifos-methyl (0.95 THz) and water (1 THz), suggesting that 2D GNR arrays are excellent platforms for the design of novel biosensing devices.

Although this modeling approach has some restrictions, such as its frequency range of use and its failure to explicitly consider the effects of sample passivation, ribbon edges, and disorder, it still demonstrates good agreement with previous experiments and predictions. This semi-analytical model is therefore useful as a preliminary step in determining the theoretical electronic and plasmonic properties of viable samples of graphene or graphene-related ribbons.

Supplementary Materials: The following supporting information can be downloaded at: https://www.mdpi.com/article/10.3390/cryst13020311/s1. Figures S1–S6.

Author Contributions: Conceptualization, T.T. and C.V.G.; Methodology, N.B.G., M.B. and C.V.G.; Validation, T.T., M.G. and C.V.G.; Formal Analysis, C.V.G.; Resources, T.T.; Data Curation, T.T.; Writing—Original Draft Preparation, C.V.G.; Writing—Review & Editing, C.V.G.; Visualization, M.G.; Project Administration, T.T.; Funding Acquisition, T.T. All authors have read and agreed to the published version of the manuscript.

Funding: This work was supported by Universidad Técnica Particular de Loja (UTPL-Ecuador) (RUC No. 1190068729001).

Acknowledgments: T.T., M.G. and C.V.G. wish to thank the Ecuadorian National Department of Sciences and Technology (SENESCYT).

Conflicts of Interest: The authors declare no conflict of interest.

References

1. Abergel, D.; Apalkov, V.; Berashevich, J.; Ziegler, K.; Chakraborty, T. Properties of graphene: A theoretical perspective. *Adv. Phys.* **2010**, *59*, 261–482. [CrossRef]
2. Castro, A.H.; Guinea, F.; Peres, N.M.R.; Novoselov, K.S.; Geim, A.K. The electronic properties of graphene. *Rev. Mod. Phys.* **2009**, *81*, 109. [CrossRef]
3. Celis, A.; Nair, M.N.; Taleb-Ibrahimi, A.; Conrad, E.H.; Berger, C.; de Heer, W.A.; Tejeda, A. Graphene nanoribbons: Fabrication, properties and devices. *J. Phys. D Appl. Phys.* **2016**, *49*, 143001. [CrossRef]
4. Novoselov, K.S.; Geim, A.K.; Morozov, S.V.; Jiang, D.; Zhang, Y.; Dubonos, S.V.; Grigorieva, I.V.; Firsov, A.A. Electric field effect in atomically thin carbon films. *Science* **2004**, *306*, 666–669. [CrossRef] [PubMed]
5. Wu, J.; Lin, H.; Moss, D.J.; Loh, K.P.; Jia, B. Graphene oxide for photonics, electronics and optoelectronics. *Nat. Rev. Chem.* **2023**, 1–22. [CrossRef]
6. Kim, K.; Choi, J.-Y.; Kim, T.; Cho, S.-H.; Chung, H.-J. A role for graphene in silicon-based semiconductor devices. *Nature* **2011**, *479*, 338–344. [CrossRef] [PubMed]
7. Kamedulski, P.; Lukaszewicz, J.; Witczak, L.; Szroeder, P.; Ziolkowski, P. The Importance of Structural Factors for the Electrochemical Performance of Graphene/Carbon Nanotube/Melamine Powders towards the Catalytic Activity of Oxygen Reduction Reaction. *Materials* **2021**, *14*, 2448. [CrossRef] [PubMed]
8. González, Z.; Acevedo, B.; Predeanu, G.; Axinte, S.M.; Drăgoescu, M.-F.; Slăvescu, V.; Fernandez, J.J.; Granda, M.; Gryglewicz, G.; Melendi-Espina, S. Graphene materials from microwave-derived carbon precursors. *Fuel Process. Technol.* **2021**, *217*, 106803. [CrossRef]
9. Tene, T.; Arias, F.A.; Guevara, M.; Nuñez, A.; Villamagua, L.; Tapia, C.; Pisarra, M.; Torres, F.J.; Caputi, L.S.; Gomez, C.V. Removal of mercury(II) from aqueous solution by partially reduced graphene oxide. *Sci. Rep.* **2022**, *12*, 6326. [CrossRef]
10. Pereira, P.; Ferreira, D.P.; Araújo, J.C.; Ferreira, A.; Fangueiro, R. The Potential of Graphene Nanoplatelets in the Development of Smart and Multifunctional Ecocomposites. *Polymers* **2020**, *12*, 2189. [CrossRef] [PubMed]
11. Li, X.; Shen, Q.; Zhang, Y.; Wang, L.; Nie, C. Wear behavior of electrodeposited nickel/graphene composite coating. *Diam. Relat. Mater.* **2021**, *119*, 108589. [CrossRef]
12. Tseluikin, V.; Zhang, L. Carbon and Carbon-Based Composite Thin Films/Coatings: Synthesis, Properties and Applications. *Coatings* **2022**, *12*, 907. [CrossRef]
13. Sahoo, S.; Bhuyan, M.; Sahoo, D. Tuning of dielectric and magnetic performance of graphene oxide via defect regulation by metal oxide nanoparticle for high temperature device. *J. Alloy. Compd.* **2023**, *935*, 168097. [CrossRef]
14. Dong, X.; Huang, W.; Chen, P. In Situ Synthesis of Reduced Graphene Oxide and Gold Nanocomposites for Nanoelectronics and Biosensing. *Nanoscale Res. Lett.* **2011**, *6*, 1–6. [CrossRef]
15. Kim, H.W.; Ko, W.; Joo, W.-J.; Cho, Y.; Oh, Y.; Ku, J.; Jeon, I.; Park, S.; Hwang, S.W. Unraveling the Structural and Electronic Properties of Graphene/Ge(110). *J. Phys. Chem. Lett.* **2018**, *9*, 7059–7063. [CrossRef] [PubMed]
16. Tian, C.; Miao, W.; Zhao, L.; Wang, J. Graphene nanoribbons: Current status and challenges as quasi-one-dimensional nanomaterials. *Rev. Phys.* **2023**, *10*, 100082. [CrossRef]
17. Yang, Y.; Murali, R. Impact of Size Effect on Graphene Nanoribbon Transport. *IEEE Electron Device Lett.* **2010**, *31*, 237–239. [CrossRef]

18. Fei, Z.; Goldflam, M.D.; Wu, J.-S.; Dai, S.; Wagner, M.; McLeod, A.S.; Liu, M.K.; Post, K.W.; Zhu, S.; Janssen, G.C.A.M.; et al. Edge and Surface Plasmons in Graphene Nanoribbons. *Nano Lett.* **2015**, *15*, 8271–8276. [CrossRef]
19. Gomez, C.V.; Pisarra, M.; Gravina, M.; Pitarke, J.M.; Sindona, A. Plasmon Modes of Graphene Nanoribbons with Periodic Planar Arrangements. *Phys. Rev. Lett.* **2016**, *117*, 116801. [CrossRef]
20. Yin, H.; Zhang, H. Plasmons in graphene nanostructures. *J. Appl. Phys.* **2012**, *111*, 103502. [CrossRef]
21. Mousavi, S.T.; Badehian, H.A.; Gharbavi, K. Optical spectra of zigzag graphene nanoribbons: A first-principles study. *Phys. Scr.* **2022**, *97*, 105803. [CrossRef]
22. Toudert, J.; Serna, R. Interband transitions in semi-metals, semiconductors, and topological insulators: A new driving force for plasmonics and nanophotonics [Invited]. *Opt. Mater. Express* **2017**, *7*, 2299–2325. [CrossRef]
23. Gan, C.H.; Gbur, G.; Visser, T.D. Surface Plasmons Modulate the Spatial Coherence of Light in Young's Interference Experiment. *Phys. Rev. Lett.* **2007**, *98*, 043908. [CrossRef] [PubMed]
24. Low, T.; Avouris, P. Graphene Plasmonics for Terahertz to Mid-Infrared Applications. *ACS Nano* **2014**, *8*, 1086–1101. [CrossRef] [PubMed]
25. Wu, L.; Chu, H.S.; Koh, W.S.; Li, E.P. Highly sensitive graphene biosensors based on surface plasmon resonance. *Opt. Express* **2010**, *18*, 14395–14400. [CrossRef]
26. Fu, H.; Zhang, S.; Chen, H.; Weng, J. Graphene Enhances the Sensitivity of Fiber-Optic Surface Plasmon Resonance Biosensor. *IEEE Sens. J.* **2015**, *15*, 5478–5482. [CrossRef]
27. Sadeghi, Z.; Shirkani, H. Highly sensitive mid-infrared SPR biosensor for a wide range of biomolecules and biological cells based on graphene-gold grating. *Phys. E. Low-dimensional Syst. Nanostructures* **2020**, *119*, 114005. [CrossRef]
28. Zhang, H.; Song, D.; Gao, S.; Zhang, J.; Zhang, H.; Sun, Y. Novel SPR biosensors based on metal nanoparticles decorated with graphene for immunoassay. *Sens. Actuators B Chem.* **2013**, *188*, 548–554. [CrossRef]
29. Mulpur, P.; Podila, R.; Lingam, K.; Vemula, S.K.; Ramamurthy, S.S.; Kamisetti, V.; Rao, A.M. Amplification of Surface Plasmon Coupled Emission from Graphene–Ag Hybrid Films. *J. Phys. Chem. C.* **2013**, *117*, 17205–17210. [CrossRef]
30. Ling, X.; Xie, L.; Fang, Y.; Xu, H.; Zhang, H.; Kong, J.; Dresselhaus, M.S.; Zhang, J.; Liu, Z. Can Graphene Be Used as a Substrate for Raman Enhancement? *Nano Lett.* **2010**, *10*, 553–561. [CrossRef]
31. Pisarra, M.; Gomez, C.V.; Sindona, A. Massive and massless plasmons in germanene nanosheets. *Sci. Rep.* **2022**, *12*, 18624. [CrossRef] [PubMed]
32. Sindona, A.; Gomez, C.V.; Pisarra, M. Dielectric screening versus geometry deformation in two-dimensional allotropes of silicon and germanium. *Sci. Rep.* **2022**, *12*, 15107. [CrossRef] [PubMed]
33. Popov, V.V.; Bagaeva, T.Y.; Otsuji, T.; Ryzhii, V. Oblique terahertz plasmons in graphene nanoribbon arrays. *Phys. Rev. B* **2010**, *81*, 73404. [CrossRef]
34. Barrios, C.A. Modeling of a Graphene Nanoribbon–based Microfluidic Surface Plasmon Resonance Biosensor. *Plasmonics* **2021**, *17*, 745–752. [CrossRef]
35. Son, J.G.; Son, M.; Moon, K.-J.; Lee, B.H.; Myoung, J.-M.; Strano, M.S.; Ham, M.-H.; Ross, C.A. Sub-10 nm Graphene Nanoribbon Array Field-Effect Transistors Fabricated by Block Copolymer Lithography. *Adv. Mater.* **2013**, *25*, 4723–4728. [CrossRef]
36. Gonze, X.; Amadon, B.; Anglade, P.-M.; Beuken, J.-M.; Bottin, F.; Boulanger, P.; Bruneval, F.; Caliste, D.; Caracas, R.; Côté, M.; et al. ABINIT: First-principles approach to material and nanosystem properties. *Comput. Phys. Commun.* **2009**, *180*, 2582–2615. [CrossRef]
37. Kohn, W. Density-functional theory for excited states in a quasi-local-density approximation. *Phys. Rev. A* **1986**, *34*, 737–741. [CrossRef]
38. Troullier, N.; Martins, J.L. Efficient Pseudopotentials for Plane-Wave Calculations. *Phys. Rev. B.* **1991**, *43*, 1993. [CrossRef]
39. Monkhorst, H.J.; Pack, J.D. Special points for Brillouin-zone integrations. *Phys. Rev. B* **1976**, *13*, 5188. [CrossRef]
40. Geim, A.K.; Novoselov, K.S. The Rise of Graphene. In *Nanoscience and Technology: A Collection of Reviews from Nature Journals*; World Scientific: Singapore, 2010; pp. 11–19.
41. Barone, V.; Hod, O.; Scuseria, G.E. Electronic Structure and Stability of Semiconducting Graphene Nanoribbons. *Nano Lett.* **2006**, *6*, 2748–2754. [CrossRef]
42. Han, M.Y.; Oezyilmaz, B.; Zhang, Y.; Kim, P. Energy Band-Gap Engineering of Graphene Nanoribbons. *Phys. Rev. Lett.* **2007**, *98*, 206805. [CrossRef]
43. Hwang, C.; Siegel, D.A.; Mo, S.-K.; Regan, W.; Ismach, A.; Zhang, Y.; Zettl, A.; Lanzara, A. Fermi velocity engineering in graphene by substrate modification. *Sci. Rep.* **2012**, *2*, 1–4. [CrossRef]
44. Zur, A.; McGill, T.C.; Smith, D.L. Fermi-level position at a semiconductor-metal interface. *Phys. Rev. B* **1983**, *28*, 2060–2067. [CrossRef]
45. Whelan, P.R.; Shen, Q.; Zhou, B.; Serrano, I.G.; Kamalakar, M.V.; Mackenzie, D.M.A.; Ji, J.; Huang, D.; Shi, H.; Luo, D.; et al. Fermi velocity renormalization in graphene probed by terahertz time-domain spectroscopy. *2D Mater.* **2020**, *7*, 035009. [CrossRef]
46. Pisarra, M.; Sindona, A.; Riccardi, P.; Silkin, V.; Pitarke, J.M. Acoustic plasmons in extrinsic free-standing graphene. *N. J. Phys.* **2014**, *16*, 83003. [CrossRef]
47. Tene, T.; Guevara, M.; Svozilík, J.; Coello-Fiallos, D.; Briceño, J.; Gomez, C.V. Proving Surface Plasmons in Graphene Nanoribbons Organized as 2D Periodic Arrays and Potential Applications in Biosensors. *Chemosensors* **2022**, *10*, 514. [CrossRef]

48. Petrakis, L. Spectral line shapes: Gaussian and Lorentzian functions in magnetic resonance. *J. Chem. Educ.* **1967**, *44*, 432. [CrossRef]
49. Blaber, M.G.; Henry, A.-I.; Bingham, J.M.; Schatz, G.C.; Van Duyne, R.P. LSPR Imaging of Silver Triangular Nanoprisms: Correlating Scattering with Structure Using Electrodynamics for Plasmon Lifetime Analysis. *J. Phys. Chem. C* **2011**, *116*, 393–403. [CrossRef]
50. Egerton, R.F. Electron energy-loss spectroscopy in the TEM. *Rep. Prog. Phys.* **2008**, *72*, 16502. [CrossRef]
51. Yang, L.; Park, C.-H.; Son, Y.-W.; Cohen, M.L.; Louie, S.G. Quasiparticle Energies and Band Gaps in Graphene Nanoribbons. *Phys. Rev. Lett.* **2007**, *99*, 186801. [CrossRef]
52. Ullah, Z.; Witjaksono, G.; Nawi, I.; Tansu, N.; Khattak, M.I.; Junaid, M. A Review on the Development of Tunable Graphene Nanoantennas for Terahertz Optoelectronic and Plasmonic Applications. *Sensors* **2020**, *20*, 1401. [CrossRef]
53. Kiraly, B.; Mannix, A.J.; Jacobberger, R.M.; Fisher, B.L.; Arnold, M.S.; Hersam, M.C.; Guisinger, N.P. Sub-5 Nm, Globally Aligned Graphene Nanoribbons on Ge (001). *Appl. Phys. Lett.* **2016**, *108*, 213101. [CrossRef]
54. Ruffieux, P.; Ruffieux, P.; Cai, J.; Cai, J.; Plumb, N.C.; Plumb, N.C.; Patthey, L.; Patthey, L.; Prezzi, D.; Prezzi, D.; et al. Electronic Structure of Atomically Precise Graphene Nanoribbons. *ACS Nano* **2012**, *6*, 6930–6935. [CrossRef]
55. Elias, D.C.; Gorbachev, R.V.; Mayorov, A.S.; Morozov, S.V.; Zhukov, A.A.; Blake, P.; Ponomarenko, L.A.; Grigorieva, I.V.; Novoselov, K.S.; Guinea, F.; et al. Dirac cones reshaped by interaction effects in suspended graphene. *Nat. Phys.* **2011**, *7*, 701–704. [CrossRef]
56. Sarker, D.; Nakti, P.P.; Tahmid, I.; Mamun, A.Z.; Zubair, A. Terahertz polarizer based on tunable surface plasmon in graphene nanoribbon. *Opt. Express* **2021**, *29*, 42713. [CrossRef]
57. Ju, L.; Geng, B.; Horng, J.; Girit, Ç.; Martin, M.; Hao, Z.; Bechtel, H.A.; Liang, X.; Zettl, A.; Shen, Y.R.; et al. Graphene plasmonics for tunable terahertz metamaterials. *Nat. Nanotechnol.* **2011**, *6*, 630–634. [CrossRef] [PubMed]
58. Xu, W.; Xie, L.; Zhu, J.; Wang, W.; Ye, Z.; Ma, Y.; Tsai, C.-Y.; Chen, S.; Ying, Y. Terahertz sensing of chlorpyrifos-methyl using metamaterials. *Food Chem.* **2017**, *218*, 330–334. [CrossRef] [PubMed]
59. Hua, X.; Qian, G.; Yang, J.; Hu, B.; Fan, J.; Qin, N.; Li, G.; Wang, Y.; Liu, F. Development of an immunochromatographic assay for the rapid detection of chlorpyrifos-methyl in water samples. *Biosens. Bioelectron.* **2010**, *26*, 189–194. [CrossRef]
60. Xu, W.; Xie, L.; Zhu, J.; Tang, L.; Singh, R.; Wang, C.; Ma, Y.; Chen, H.-T.; Ying, Y. Terahertz biosensing with a graphene-metamaterial heterostructure platform. *Carbon* **2018**, *141*, 247–252. [CrossRef]
61. Zhou, J.; Zhao, X.; Huang, G.; Yang, X.; Zhang, Y.; Zhan, X., Tian, H.; Xiong, Y.; Wang, Y.; Fu, W. Molecule-Specific Terahertz Biosensors Based on an Aptamer Hydrogel-Functionalized Metamaterial for Sensitive Assays in Aqueous Environments. *ACS Sensors* **2021**, *6*, 1884–1890. [CrossRef]
62. Lin, T.N.; Chih, K.H.; Yuan, C.T.; Shen, J.L.; Lin, C.A.J.; Liu, W.R. Laser-ablation production of graphene oxide nanostructures: From ribbons to quantum dots. *Nanoscale* **2015**, *7*, 2708–2715. [CrossRef] [PubMed]

Disclaimer/Publisher's Note: The statements, opinions and data contained in all publications are solely those of the individual author(s) and contributor(s) and not of MDPI and/or the editor(s). MDPI and/or the editor(s) disclaim responsibility for any injury to people or property resulting from any ideas, methods, instructions or products referred to in the content.

crystals

MDPI

Article

Cyclic Voltammetry and Impedance Measurements of Graphene Oxide Thin Films Dip-Coated on n-Type and p-Type Silicon

Grazia Giuseppina Politano [1,*], Stefano Burza [1] and Carlo Versace [1,2]

[1] Dipartimento di Fisica, Università della Calabria, 87036 Rende, CS, Italy
[2] Licryl CNR/Nanotec c/o Dipartimento di Fisica, Università della Calabria, 87036 Rende, CS, Italy
* Correspondence: grazia.politano@unical.it

Abstract: Despite the increasing interest in graphene, a less studied aspect is the enhancement of silicon (Si) performances due to the interaction with graphene-based materials. In this study, cyclic voltammetry and electric impedance measurements are performed on graphene oxide (GO) dip-coated on n-type and p-type Si samples. The electrical properties of GO on n-type Si samples are dramatically enhanced: The conductivity and the photocurrent meaningfully increase in comparison to bare n-type Si. Such findings could be used in a wide variety of optoelectronic applications, improving GO future applicability in the Si semiconductor industry.

Keywords: graphene oxide; silicon; thin films; impedance; cyclic voltammetry

1. Introduction

The semiconductor technology is based on crystalline silicon (Si), which is an inefficient light-emitting material because it is an indirect bandgap semiconductor [1]. Therefore, attention in the semiconductor field has turned to graphene, which is a single layer of carbon atoms arranged in a honeycomb lattice [2] that has ultrahigh charge-carrier mobility, broadband light absorption, and linear dispersion band structure [3]. Graphene is thus an advantageous material for several applications in electronics and optoelectronics [4,5].

Despite its outstanding properties, graphene cannot replace Si entirely because of the poor on/off current ratio, which results from its zero bandgap [6]. However, it has been suggested that graphene could improve Si-based devices to afford broadband operation in light routing and amplification in components such as polarizers, modulators, and photodetectors [6].

Another advantage of graphene-based materials is that they can be produced in ultrathin sheet form that can be transferred to suitable substrates [7].

Graphene oxide (GO) is a graphene-based material that has more oxygen-containing groups and defects compared with mechanically exfoliated or chemical vapor-deposited graphene. These defects are proven to be favorable to enhancing the performance of photodetectors [8]. Furthermore, previous research works have shown that GO thin films have high optical transmittance in the visible region of the spectrum that allows their use as protective coatings and optically transparent electrodes, crucial in solar cells and for optical applications [9].

GO has been studied for several applications [10–24] in previous research works.

In particular, Brzhezinskaya et al. [25] demonstrated a simple method for GO deposition for creating different electronic devices including memristor devices for neuromorphic computing systems in the field of large data and artificial intelligence. Khurana et al. [26] observed thermally stable bipolar resistive switching behavior in the memory devices composed of GO thin films on indium tin oxide coated glass substrate with platinum as the top electrode. Panin et al. [27] reported resistive switching behaviors in an Al/GO/Al planar structure. Qi et al. [28] demonstrated highly uniform resistive switching and a high on/off ratio of resistive random access memory based on GO by embedding gold nanoparticles into the device. Jeong et al. [28] presented a GO-based memory that can be easily fabricated using a room-temperature spin-casting method on flexible substrates and has reliable memory performance in terms of retention and endurance.

In our works, we have studied the optical properties of GO [9,29]. A less studied aspect in the framework of graphene-based material is the enhancement of Si performances due to the interaction with GO.

In this work, we characterize GO thin films deposited on n-type and p-type Si substrates using cyclic voltammetry and impedance measurements. The photocurrent and conductivity of n-type Si are substantially enhanced due to the interaction with GO thin films, which could be crucial for several applications in the optical field.

2. Materials and Methods

The n-type and p-type Si wafers used as substrates were ultrasonically cleaned in acetone, then in bidistilled water, and finally in isopropanol.

All samples were prepared using GO 2 g/L dispersion in H_2O (bought from Punto Quantico).

The dip-coating process [30] was used to deposit GO films (10 nm) on n-type and p-type Si with a homemade apparatus at a speed of 0.33 mm/s.

In Figures 1 and 2, the schemes of the sample cell are reported.

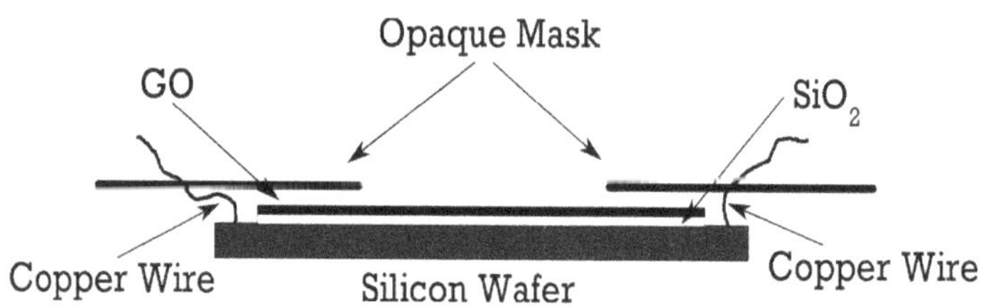

Figure 1. Scheme of the sample cell.

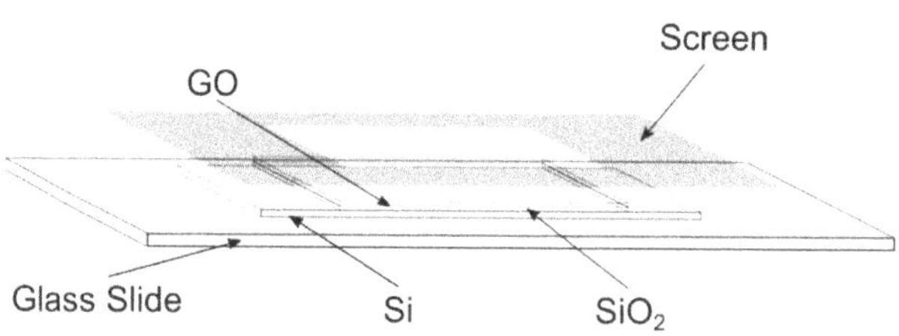

Figure 2. Scheme of the sample cell.

The ends of the silicon strip were etched with a hydrogen fluoride solution to remove the native SiO_2 layer, and two copper conductors were indium soldered on both ends. An opaque mask reduced the light-exposed surface of the sample to an area of 17 mm × 5 mm.

In Figure 4, the scheme of the experimental setup is reported.

In Figure 3, the photos of the sample cell are reported.

The sample was placed inside an insulated dark box, the temperature of which was controlled by a fan and a thermometer. The light beam was emitted by a tungsten halogen lamp powered at a constant current of 8.3A (100 W) using an Instek 2010 PSP-2010 power supply. Current-to-voltage scans were performed using a Keithley 2612A source meter. Sample impedance was measured using an Agilent 4294A Precision Impedance Analyzer. All the measurements were made at room temperature and were performed inside the dark box in conditions of dark and light illumination (4450 lux).

Additional morphological, optical, and electrical measurements of the samples are available in the Supplementary Materials.

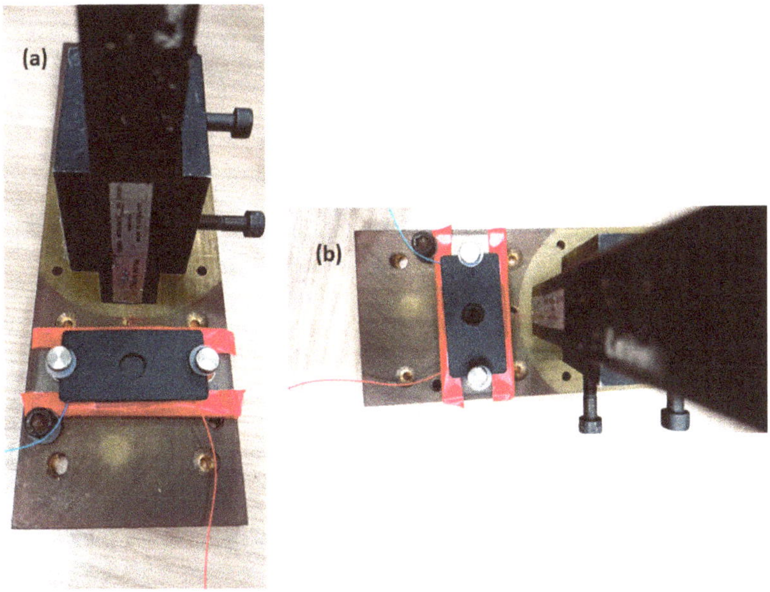

Figure 3. Photos of the sample cell. (**a**) front view (**b**) top view.

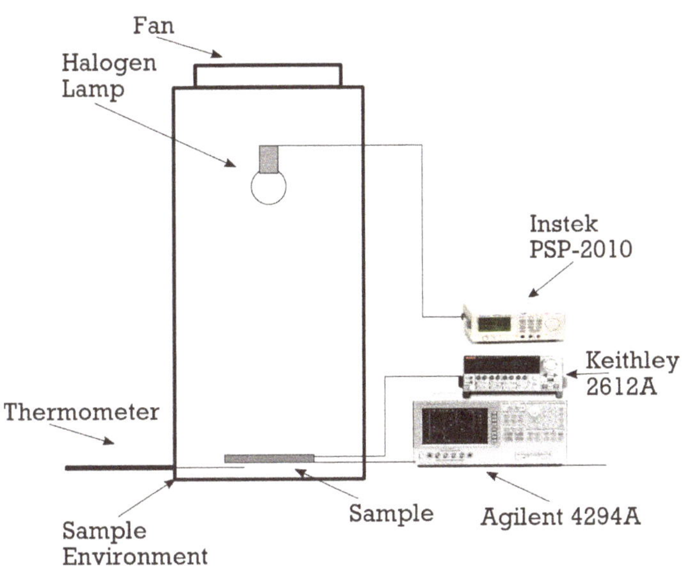

Figure 4. Experimental setup.

3. Results and Discussion

In Figures 5 and 6, the cyclic voltammetry measurements of n-type (p-type) Si and GO/n-type (p-type) Si samples under dark and light illumination conditions are reported.

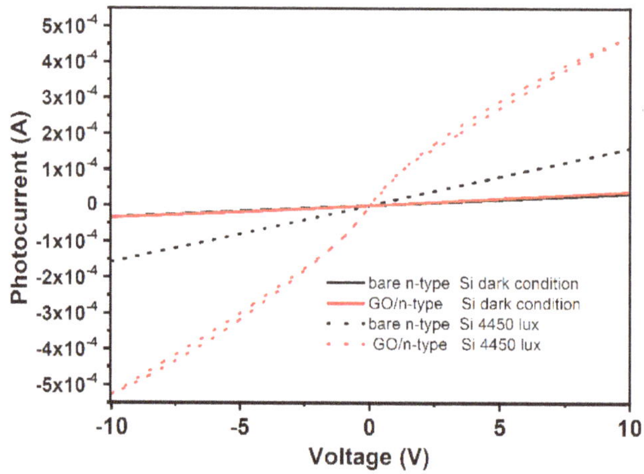

Figure 5. Cyclic voltammetry measurements on n-type silicon and graphene oxide/n-type Si samples under dark and light illumination conditions.

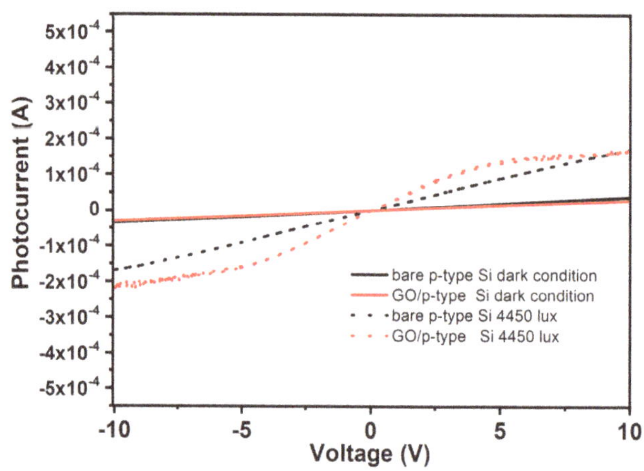

Figure 6. Cyclic voltammetry measurements on p-type silicon and graphene oxide/p-type Si samples under dark and light illumination conditions.

As can be seen in Figure 5, the I–V characteristics show that the photocurrent of the GO/n-type Si samples more than tripled in comparison to bare n-type Si samples under light illumination.

Under light illumination conditions, incident light is absorbed in the Si substrate and electron−hole pairs are generated as a result. Therefore, an explanation of the observed behavior may be that photogenerated carriers in GO can be driven toward the Si conduction band by a built-in electrical field with light illumination. At the same time, Si also absorbs photons and excites the photogenerated carriers. Electrons will accumulate in the conduction band of Si, while photogenerated holes will be transported into the GO's valence band. The mechanism of generation and transportation inhibits the effective recombination of carriers, and in turn, increases the photocurrent.

The I–V characteristics in Figure 6 reveal a current behavior in the GO/p-type Si samples almost identical to that of the bare p-type Si in dark and light illumination conditions. Therefore, there is no substantial improvement due to the presence of GO in the case of p-type Si substrates.

Figures 7 and 8 show the impedance measurements on n-type (p-type) Si and GO/n-type (p-type) Si samples.

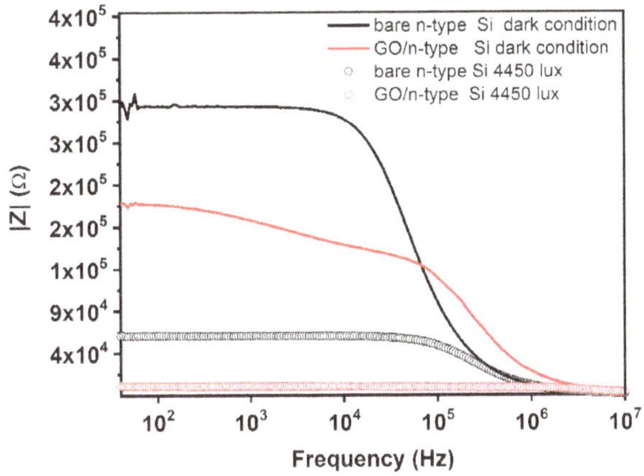

Figure 7. Impedance measurements on n-type silicon and graphene oxide/n-type Si samples under dark and light illumination conditions.

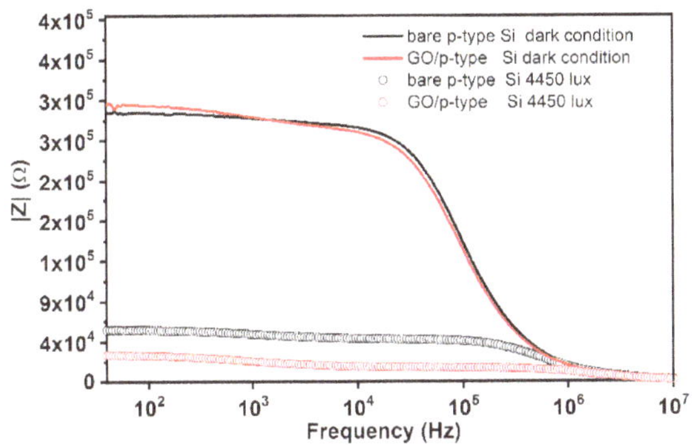

Figure 8. Impedance measurements on p-type silicon and graphene oxide/p-type Si samples under dark and light illumination conditions.

In Figure 7, the impedance of GO/n-type Si samples decreases in both the absence and presence of light radiation. The conductivity is quintupled for GO/n-type Si samples in comparison to bare n-type Si under light conditions.

A possible explanation may be that GO doping shifts the Fermi energy level of Si towards the Si conduction band, thus increasing its conductivity. However, additional research is required to corroborate this claim.

As shown in Figure 8, the p-type silicon and GO/p-type Si samples have the same impedance behavior in dark conditions. In the presence of light radiation, there is an increase in GO/p -type Si conductivity in comparison to bare p-type Si.

4. Conclusions

In summary, this study provides a demonstration of an innovative method for improving the conductivity of Si using GO.

The photocurrent and conductivity of n-type Si are significantly enhanced by dip-coating GO on it. The photocurrent of p-type Si is not significantly affected by interaction with GO, whereas its conductivity is enhanced.

Hence, GO could improve the Si semiconductor industry, (especially n-type Si), finding applications in numerous areas such as solar cells, touch screens, transparent displays, imaging, sensing, and other optoelectronic applications.

Supplementary Materials: The following supporting information can be downloaded at: https://www.mdpi.com/article/10.3390/cryst13010073/s1, Figure S1: Scanning electron microscopy image of GO films on Si substrates (n-type). Figure S2: Scanning electron microscopy image of GO films on Si substrates (p-type). Figure S3: Particle size distribution of very diluted GO flakes in water. Figure S4: Variable angle spectroscopic ellipsometry measurements of graphene oxide films on Si substrates (n-type). Experimental and model generated ψ (a) and Δ (b) data fit at different angles of incidence. Figure S5: Variable angle spectroscopic ellipsometry measurements of graphene oxide films on Si substrates (p-type). Experimental and model generated ψ (a) and Δ (b) data fit at different angles of incidence. Figure S6: Impedance measurements on n-type Si and graphene oxide/n-type Si samples under dark and light illumination conditions. Figure S7: Cyclic voltammetry measurements on n-type Si and graphene oxide/n-type Si samples under dark and light illumination conditions.

Author Contributions: Investigation, G.G.P., S.B. and C.V.; supervision, C.V.; writing—original draft preparation, G.G.P. and C.V.; writing—review and editing, G.G.P. and C.V. All authors have read and agreed to the published version of the manuscript.

Funding: This research received no external funding.

Data Availability Statement: Not applicable.

Conflicts of Interest: The authors declare no conflict of interest.

References

1. Hirschman, K.D.; Tsybeskov, L.; Duttagupta, S.P.; Fauchet, P.M. Silicon-based visible light-emitting devices integrated into microelectronic circuits. *Nature* **1996**, *384*, 338. [CrossRef]
2. Geim, A.K.; Novoselov, K.S. The rise of graphene. *Nat. Mater.* **2007**, *6*, 183–191. [CrossRef]
3. Castriota, M.; Politano, G.G.; Vena, C.; De Santo, M.P.; Desiderio, G.; Davoli, M.; Cazzanelli, E.; Versace, C. Variable Angle Spectroscopic Ellipsometry investigation of CVD-grown monolayer graphene. *Appl. Surf. Sci.* **2019**, *467–468*, 213–220. [CrossRef]
4. Hsieh, Y.-P.; Yen, C.-H.; Lin, P.-S.; Ma, S.-W.; Ting, C.-C.; Wu, C.-I.; Hofmann, M. Ultra-high sensitivity graphene photosensors. *Appl. Phys. Lett.* **2014**, *104*, 41110. [CrossRef]
5. Callicó, G.M. Image sensors go broadband. *Nat. Photonics* **2017**, *11*, 332. [CrossRef]
6. Kim, K.; Choi, J.-Y.; Kim, T.; Cho, S.-H.; Chung, H.-J. A role for graphene in silicon-based semiconductor devices. *Nature* **2011**, *479*, 338. [CrossRef]
7. Yi, M.; Shen, Z. A review on mechanical exfoliation for the scalable production of graphene. *J. Mater. Chem. A* **2015**, *3*, 11700–11715. [CrossRef]
8. Zhu, M.; Li, X.; Guo, Y.; Li, X.; Sun, P.; Zang, X.; Wang, K.; Zhong, M.; Wu, D.; Zhu, H. Vertical junction photodetectors based on reduced graphene oxide/silicon Schottky diodes. *Nanoscale* **2014**, *6*, 4909–4914. [CrossRef]
9. Politano, G.G.; Versace, C.; Vena, C.; Castriota, M.; Ciuchi, F.; Fasanella, A.; Desiderio, G.; Cazzanelli, E. Physical investigation of electrophoretically deposited graphene oxide and reduced graphene oxide thin films. *J. Appl. Phys.* **2016**, *120*, 195307. [CrossRef]
10. Nguyen, Q.K.; Kuz'micheva, G.M.; Khramov, E.V.; Svetogorov, R.D.; Chumakov, R.G.; Cao, T.T. Design of Metal-Organic Polymers MIL-53(M3+): Preparation and Characterization of MIL-53(Fe) and Graphene Oxide Composite. *Crystals* **2021**, *11*, 1281. [CrossRef]
11. Perumal, D.; Albert, E.L.; Saad, N.; Hin, T.Y.; Zawawi, R.M.; Teh, H.F.; Che Abdullah, C.A. Fabrication and Characterization of Clinacanthus nutans Mediated Reduced Graphene Oxide Using a Green Approach. *Crystals* **2022**, *12*, 1539. [CrossRef]
12. Zhou, S.; Zhang, X.; Zhou, H.; Li, D. Effects of Graphene Oxide Encapsulated Silica Fume and Its Mixing with Nano-Silica Sol on Properties of Fly Ash-Mixed Cement Composites. *Crystals* **2022**, *12*, 144. [CrossRef]
13. Politano, G.G.; Versace, C. Variable Angle Spectroscopic Ellipsometry Characterization of Graphene Oxide in Methanol Films. *Crystals* **2022**, *12*, 696. [CrossRef]
14. Wang, X.; Lu, S.; Xu, W. Synthesis of Needle-like Nanostructure Composite Electrode of Co_3O_4/rGO/NF for High-Performance Symmetric Supercapacitor. *Crystals* **2022**, *12*, 664. [CrossRef]
15. Alshoaibi, A.; Awada, C.; Ahmed, F.; Obodo, R.M.; Maaza, M.; Ezema, F.I. Trimetallic Oxides/GO Composites Optimized with Carbon Ions Radiations for Supercapacitive Electrodes. *Crystals* **2022**, *12*, 874. [CrossRef]
16. Wang, S.; Wang, X.; He, J.; Xin, M. Mechanical Behavior and Microstructure of Graphene Oxide Electrodeposited Carbon Fiber Reinforced Cement-Based Materials. *Crystals* **2022**, *12*, 964. [CrossRef]
17. Abo-Dief, H.M.; Alanazi, A.K.; Alothman, Z.A.; Pramanik, T.; Mohamed, A.T.; Fallata, A.M.; Althakafy, J.T. An Investigation into the Production of rGO/CuO Composites Using Plant Wastes. *Crystals* **2022**, *12*, 1423. [CrossRef]

18. Alanazi, A.K.; Abo-Dief, H.M.; Alothman, Z.A.; Mohamed, A.T.; Pramanik, T.; Fallata, A.M. Effect of rGO wt.% on the Preparation of rGO/CuO Nanocomposites at Different Test Periods and Temperatures. *Crystals* **2022**, *12*, 1325. [CrossRef]
19. Kashif, M.; Alsaiari, N.S.; Jaafar, E.; Low, F.W.; Oon, C.S.; Sahari, S.K.; Almuaikel, N.S. Reaction-Time-Dependent Opto-Electrical Properties of Graphene Oxide. *Crystals* **2022**, *12*, 1303. [CrossRef]
20. Politano, G.G.; Versace, C. Electrical and Optical Characterization of Graphene Oxide and Reduced Graphene Oxide Thin Films. *Crystals* **2022**, *12*, 1312. [CrossRef]
21. Xu, X.; Xiong, C.; Mao, S.; Yao, W. Established Model on Polycrystalline Graphene Oxide and Analysis of Mechanical Characteristic. *Crystals* **2022**, *12*, 382. [CrossRef]
22. Ansari, A.R.; Ansari, S.A.; Parveen, N.; Ansari, M.O.; Osman, Z. Silver Nanoparticle Decorated on Reduced Graphene Oxide-Wrapped Manganese Oxide Nanorods as Electrode Materials for High-Performance Electrochemical Devices. *Crystals* **2022**, *12*, 389. [CrossRef]
23. Riley, P.R.; Joshi, P.; Penchev, H.; Narayan, J.; Narayan, R.J. One-Step Formation of Reduced Graphene Oxide from Insulating Polymers Induced by Laser Writing Method. *Crystals* **2021**, *11*, 1308. [CrossRef]
24. Shireesha, K.; Kumar, T.R.; Rajani, T.; Chakra, C.S.; Kumari, M.M.; Divya, V.; Raghava Reddy, K. Novel NiMgOH-rGO-Based Nanostructured Hybrids for Electrochemical Energy Storage Supercapacitor Applications: Effect of Reducing Agents. *Crystals* **2021**, *11*, 1144. [CrossRef]
25. Brzhezinskaya, M.; Kapitanova, O.O.; Kononenko, O.V.; Koveshnikov, S.; Korepanov, V.; Roshchupkin, D. Large-scalable graphene oxide films with resistive switching for non-volatile memory applications. *J. Alloys Compd.* **2020**, *849*, 156699. [CrossRef]
26. Khurana, G.; Misra, P.; Katiyar, R.S. Forming free resistive switching in graphene oxide thin film for thermally stable nonvolatile memory applications. *J. Appl. Phys.* **2013**, *114*, 124508. [CrossRef]
27. Panin, G.N.; Kapitanova, O.O.; Lee, S.W.; Baranov, A.N.; Kang, T.W. Resistive Switching in Al/Graphene Oxide/Al Structure. *Jpn. J. Appl. Phys.* **2011**, *50*, 70110. [CrossRef]
28. Qi, M.; Cao, S.; Yang, L.; You, Q.; Shi, L.; Wu, Z. Uniform multilevel switching of graphene oxide-based RRAM achieved by embedding with gold nanoparticles for image pattern recognition. *Appl. Phys. Lett.* **2020**, *116*, 163503. [CrossRef]
29. Politano, G.G.; Vena, C.; Desiderio, G.; Versace, C. Spectroscopic ellipsometry investigation of the optical properties of graphene oxide dip-coated on magnetron sputtered gold thin films. *J. Appl. Phys.* **2018**, *123*, 55303. [CrossRef]
30. Scriven, L.E. Physics and Applications of DIP Coating and Spin Coating. *MRS Proc.* **1988**, *121*, 717. [CrossRef]

Disclaimer/Publisher's Note: The statements, opinions and data contained in all publications are solely those of the individual author(s) and contributor(s) and not of MDPI and/or the editor(s). MDPI and/or the editor(s) disclaim responsibility for any injury to people or property resulting from any ideas, methods, instructions or products referred to in the content.

MDPI AG
Grosspeteranlage 5
4052 Basel
Switzerland
Tel.: +41 61 683 77 34

Crystals Editorial Office
E-mail: crystals@mdpi.com
www.mdpi.com/journal/crystals

Disclaimer/Publisher's Note: The title and front matter of this reprint are at the discretion of the Guest Editors. The publisher is not responsible for their content or any associated concerns. The statements, opinions and data contained in all individual articles are solely those of the individual Editors and contributors and not of MDPI. MDPI disclaims responsibility for any injury to people or property resulting from any ideas, methods, instructions or products referred to in the content.

www.ingramcontent.com/pod-product-compliance
Lightning Source LLC
Chambersburg PA
CBHW040717270926
51850CB00023B/2679
* 9 7 8 3 7 2 5 8 7 1 7 2 8 *